Vukota Boljanovic, Ph.D.

APPLIED MATHEMATICAL
&
PHYSICAL FORMULAS

A POCKET REFERENCE GUIDE
FOR STUDENTS, MECHANICAL ENGINEERS,
ELECTRICAL ENGINEERS, MANUFACTURING
ENGINEERS,
MAINTENANCE TECHNICIANS, TOOLMAKERS, AND
MACHINISTS

Industrial Press
New York

Library of Congress Cataloging-in-Publication Data

Boljanovic, Vukota.
  Mathematical and physical formulas pocket referenc / Vukota
Boljanovic.
     p. cm.
   ISBN 0-8311-3309-0
     1. Mathematics--Formulae--Handbooks, manuals, etc. 2.
Physics--Formulae--Handbooks, manuals, etc. I. Title
     QA41.B58 2006
     510.2'12--dc.22

                                              2006041796

Industrial Press, Inc.
989 Avenue of the Americas
New York, NY 10018

First Printing, 2007

Sponsoring Editor: John Carleo
Cover Design: Janet Romano

2   3   4   5   6   7   8   9   10

DEDICATION

To my granddaughter, *Milla Liv Broadwater*

TABLE OF CONTENTS

Preface   **xxi**

PART I **UNITS**   **1**

INTERNATIONAL SYSTEM OF UNITS   3
1. SI Base Units   3
2. SI Derived Units   3
3. SI Derived Units with Special Names and Symbols   4
4. SI Derived Units Whose Names and Symbols Include SI Derived Units with Special Names and Symbols   6
5. Units Outside the SI that Are Accepted for Use with the SI   7
6. SI Prefixes   9

METRIC UNITS OF MEASUREMENT   9
7. Units of Length   9
8. Units of Area   10
9. Units of Liquid Value   10
10. Units of Volume   11
11. Units of Mass   11

U.S. UNITS OF MEASUREMENT   12
  12. Units of Length   12
  13. Units of Area   12
  14. Units of Liquid Volume   13
  15. Units of Volume   13
  16. Apothecaries' Units of Liquid Volume   14
  17. Units of Dry Volume   14
  18. Avoirdupois Units of Mass   14
  19. Apothecaries' Units of Mass   15
  20. Troy Units of Mass   15

TABLES OF EQUIVALENTS   16
  21. Units of Length   16
  22. Units of Area   18
  23. Units of Volume   19
  24. Units of Mass   20

PART II **MATHEMATICS**   **23**

ALGEBRA   25
  1. Sets of Real Numbers   27
  2. Properties of Real Numbers   28
  3. Properties of Equality   28
  4. Properties of Fractions   29
  5. Division Properties of Zero   30
  6. Real Number Line   30
  7. Interval   31
  8. Absolute Value   32

9. Distance between Two Points on
    the Number Line   33
10. Definition of Positive Integer Exponents   33
11. Definition of b^0 34
12. Definition of b^{-n} 34
13. Properties of Exponents   34
14. Definition of $\sqrt[n]{a}$ 35
15. Properties of Radicals   35
16. General Form of a Polynomial   36
17. Factoring Polynomials   36
18. Order of Operations   37
19. Adding and Subtracting Polynomials   39
20. Multiplying Polynomials   39
21. Dividing Polynomials   39
22. Rational Expressions   40
23. Complex Fractions   43
24. Definition of a Complex Number   44
25. Definition of a Linear Equation   46
26. Addition and Multiplication Properties of
    Equality   46
27. Systems of Linear Equations   47
28. Determinants   49
29. Quadratic Equations   51
30. Properties of Inequalities   56
31. Arithmetic Sequence   56
32. Arithmetic Series   57
33. Geometric Sequences   58
34. Geometric Series   59

35. Binomial Theorem — 61
36. The Cartesian Coordinate System — 63
37. Linear Functions — 63
38. Forms of Linear Equations — 64
39. Quadratic Functions — 65
40. Basic Operation of Functions — 67
41. Exponential Functions — 67
42. Natural Exponential Function — 68
43. Logarithmic Functions — 69

GEOMETRY — 73
1. Definition of an Angle — 75
2. Unit Measurement of Angles — 75
3. Acute Angles — 75
4. Obtuse Angles — 76
5. Right Angles — 76
6. Complementary Angles — 74
7. Supplementary Angles — 76
8. Vertical Angles — 77
9. Alternate Interior Angles — 77
10. Alternate Exterior Angles — 78
11. Corresponding Angles — 78
12. Angle Bisector — 78
13. Perpendicular Angles — 79
14. Thales' Theorem — 79
15. Oblique Triangle — 79
16. Geocenter of a Triangle — 81
17. Orthocenter — 81
18. Similarity of Triangles — 81

19. The Law of Cosines   82
20. The Law of Sines   83
21. Right Triangle   83
22. Ratio of the Sides of a Right Triangle   85
23. Pythagorean Theorem   85
24. Equilateral Triangles   86
25. Isosceles Triangle   87
26. Square   88
27. Rectangle   89
28. Parallelogram   90
29. Rhombus   91
30. Trapezoid (American definition)   92
31. Kite   93
32. Regular Polygons   94
33. Circle   95
34. Sector of a Circle   96
35. Segment of a Circle   97
36. Annulus (Circular Ring)   98
37. Ellipse   98
38. Cube   100
39. Cuboid   100
40. Right Prism   101
41. Pyramid   102
42. Frustum of Pyramid   103
43. Cone   103
44. Frustum of Cone   105
45. Cylinder   106
46. Hollow Cylinder   107
47. Sliced Cylinder   108

48. Sphere   108
49. Spherical Cap   109
50. Sector of a Sphere   110
51. Zone of a Sphere   110
52. Torus   111
53. Ellipsoid   112
54. Barrel   113

TRIGONOMETRY   115
1. Circular and Angular Measures   117
2. Trigonometric Circle   118
3. Basic Formulas   119
4. Trigonometric Ratios for Right Angled
   Triangles   120
5. Sum and Difference of Functions of
   Angles   121
6. Sum and Difference of Angles   122
7. Double Angle Formulas   122
8. Half Angle Formulas   122
9. Functions of Important Angles   123
10. Solving Trigonometric Equations   124
11. Verifying Trigonometric Identities   126
12. Graphs of the Sine and Cosine Functions   128
13. Graphs of the Tangent and Cotangent
    Functions   128

ANALYTICAL GEOMETRY   131
1. Distance between Two Points   133
2. Point of Division   133

3.	Inclination and Slope of a Line	134
4.	Parallel and Perpendicular Lines	135
5.	Angle between Two Intersection Lines	136
6.	Triangle	136
7.	The Equation for a Straight Line through a Point	137
8.	Slope-Intercept Form	138
9.	Equation for a Straight Line through Two Points	138
10.	Intercept Form Equation of a Straight Line	13
11.	General Form of an Equation of a Straight Line	139
12.	Normal Equation of a Straight Line	13
13.	Distance from a Line to a Point	140
14.	Circles	141
15.	Ellipses	143
16.	Parabolas	143
17.	Hyperbolas	146
18.	Polar Coordinates	148
19.	Cartesian Coordinates	149
20.	Distance between Two Points	150
21.	Angle between Two Lines	151
22.	Every Plane	153
23.	Line Perpendicular to Plane	153
24.	Parallel and Perpendicular Planes	153
25.	Distance of a Point from a Plane	154
26.	Normal Form	154
27.	Intercept Form	155

28. Angle Between Two Planes 155
29. The Straight Line in Space 156
30. Parametric Form Equations of a Line 156
31. Symmetric Form Equations of a Line 157
32. Two Points Form Equations of a Line 158
33. Relative Directions of a Line and Plane 158
34. The Sphere 158
35. The Ellipsoid 159
36. Hyperboloid 160
37. Elliptic Paraboloid 161
38. Hyperbolic Paraboloid 162
39. Cylindrical Surface 163

MATHEMATICS OF FINANCE 165
1. Simple Interest 167
2. Compound Interest 168
3. Annuities 171
4. Amortization of Loans 174
5. Sinking Fund Payment 174

CALCULUS 177
1. Limits 179
2. Rule for Limits 179
3. Slope of Tangent Line 180
4. Definition of the Derivative 181
5. Basic Derivatives 182
6. Increasing and Decreasing Function
$y = f(x)$ 186
7. Maximum and Minimum Function

$y = f(x)$ 186
8. Solving Applied Problems 187
9. Integration 187
10. Basic Integration Rules 188
11. Integration by Substitution 189
12. Basic Integrals 191
13. Arc Length 200
14. Finding an Area and a Volume 201
15. Finding the Area between Two Curves 202

STATISTICS 205
1. Definition of Set and Notation 207
2. Terms and Symbols 208
3. Venn Diagrams 208
4. Operations on Sets 209
5. De Morgan's Laws 210
6. Counting the Elements in a Set 210
7. Permutations 211
8. Combinations 212
9. Probability Terminology 212
10. Basic Probability Principles 213
11. Random Variable 214
12. Mean Value $\bar{x}$ or Expected Value μ 214
13. Variance 215
14. Standard Deviation 215
15. Normal Distribution 216
16. Binomial Distribution 217
17. Poisson Distribution 218

18. Exponential Distribution   219
19. General Reliability Definitions   220
20. Exponential Distribution Used as
    Reliability Function   221

PART III **PHYSICS**   **223**

MECHANICS   225
 1. Scalars and Vectors   227
 2. Distance and Displacement   227
 3. Acceleration   227
 4. Speed and Velocity   228
 5. Frequency   228
 6. Period   228
 7. Angular Displacement   228
 8. Angular Velocity   228
 9. Angular Acceleration   229
10. Rotational Speed   229
11. Uniform Linear Motion   229
12. Uniform Accelerated Linear Motion   230
13. Rotational Motion   231
14. Uniform Rotation about a Fixed Axis   233
15. Uniform Accelerated Rotation about a
     Fixed Axis   234
16. Simple Harmonic Motion   235
17. Pendulum   237
18. Free Fall   238
19. Vertical Projection   239
20. Angled Projection   240

21. Horizontal Projection   241
22. Sliding Motion on an Inclined Plane   242
23. Rolling Motion on an Inclined Plane   243
24. Newton's First Law of Motion   246
25. Newton's Second Law   246
26. Newton's Third Law   246
27. Momentum of Force   247
28. Impulse of Force   247
29. Law of Conservation of Momentum   247
30. Friction   248
31. General Law of Gravity   249
32. Gravitational Force   250
33. Centrifugal Force   250
34. Centripetal Force   251
35. Torque   252
36. Work   252
37. Energy   254
38. Conservation of Energy   255
39. Relativistic Energy   256
40. Power   256
41. Resolution of a Force   257
42. Moment of a Force about a Point O   258
43. Mechanical Advantage of
    Simple Machines   258
44. The Lever   259
45. Wheel and Axle   259
46. The Pulley   260
47. The Inclined Plane   261
48. The Wedge   262

49. The Screw   263

MECHANICS OF FLUID   265
    1. Density   267
    2. Viscosity   267
    3. Pressure of Solid   268
    4. Pressure of Liquids   269
    5. Force Exerted by Liquids   270
    6. Pascal's Principle   271
    7. Archimedes' Principle   272
    8. Buoyant Force   272
    9. Flow Rate   274
   10. Conservation of Mass   274
   11. Bernoulli's Equation   275

TEMPERATURE AND HEAT   277
    1. Pressure   279
    2. Temperature   279
    3. Density   280
    4. Specific Volume   280
    5. Molar Mass   281
    6. Molar Volume   282
    7. Heat   282
    8. Specific Heat   283
    9. Heat Conduction   283
   10. Expansion of Solid Bodies   283
   11. Expansion of Liquids   284
   12. Expansion of Water   285
   13. Fusion   285

14. Vaporization 286
15. Equation of State 286
16. The Charles Law for Temperature 287
17. Boyle's Law for Pressure 287
18. Gay-Lussac's Law for Temperature 288
19. Dalton's Law of Partial Pressures 288
20. Combined Gas Law 289
21. The First Law of Thermodynamics 289
22. The Second Law of Thermodynamics 290
23. The Third Law of Thermodynamics 292

ELECTRICITY AND MAGNETISM 293
1. Coulomb's Law 295
2. Electric Fields 295
3. Electric Flux 296
4. Gauss' Law 297
5. Electric Potential 298
6. Electric Potential Energy 299
7. Capacitance 299
8. Capacitor 300
9. Electric Current 301
10. Current Density 302
11. Potential Difference 302
12. Resistance 302
13. Ohm's Law 303
14. Series Circuits 304
15. Parallel Circuits 304
16. Series-Parallel Circuit 305
17. Joule's Law 306

18. Kirchhoff's Junction Law   307
19. Kirchhoff's Loop Law   308
20. Resistors   309
21. Internal Resistance   309
22. Magnetic Forces on Moving Charges   310
23. Force on a Current-Carrying Wire   310
24. Magnetic Field of a Moving Charge   311
25. Magnetic Field of a Loop   311
26. Faraday's Law   312
27. Properties of Alternating Current   312
28. Period   313
29. Frequency   313
30. Wavelength   313
31. Instantaneous Current and Voltage   314
32. Effective Current and Voltage   314
33. Maximum Current and Voltage   315
34. Ohm's Law of AC Current Containing Only
    Resistance   315
35. AC Power   316
36. Changing Voltage with Transformers   316
37. Inductive Reactance   317
38. Inductance and Resistance in Series   318
39. Capacitance   319
40. Capacitance and Resistance in a Series   319
41. Capacitance, Inductance, and
    Resistance in Series   321
42. Power in AC Circuits   322
43. Parallel Circuit   324

LIGHT   327
   1. Visible Light   329
   2. Speed of Light   329
   3. Light as a Particle   329
   4. Luminous Intensity   329
   5. Luminous Flux   330
   6. Luminous Energy   331
   7. Illuminance   331
   8. Luminance   332
   9. Laws of Reflection   333
  10. Refraction   333
  11. Polarization   335
  12. Plane Mirrors   336
  13. Concave Mirrors   336
  14. Convex Mirrors   336
  15. Mirror Formula   337
  16. Lens Equation   337

WAVE MOTION AND SOUND   339
   1. Definition and Graph   341
   2. Wavelength   341
   3. Amplitude   342
   4. Velocity   342
   5. Frequency   342
   6. Period   342
   7. Wave on a Stretched String   343
   8. The Sinusoidal Wave   343
   9. Electromagnetic Waves   345
  10. Electromagnetic Energy   345

11. The Electromagnetic Spectrum   346
12. Sound Waves   346
13. Speed of Sound in Air   347
14. Sound Speed in Gases   347
15. The Doppler Effect   348

APPENDIX   **349**
1. Fundamental Physics Constants   349

INDEX   **353**

PREFACE

A comprehensive pocket reference guide giving students, engineers, toolmakers, metalworkers, and other specialists a wide range of mathematical and physical formulas in a handy format.

Great care has been taken to present all formulas concisely, simply, and clearly. All the information included is practical -- rarely used formulas are excluded.

Compactly arranged in an attractive, unique style, this reference book has just about every equation, definition, diagram, and formula that a user might want in doing undergraduate-level physics and mathematics.

Each year, these indispensable study guides will be a good help to hundreds of thousands of students to improve their test scores and final grades.

Thoroughly practical and authoritative, this book brings together in three parts more than a thousand formulas and figures to simplify review or to refresh your memory of what you studied in school. If you are in school now, and you don't have a lot of time but want to excel in class, this book will help you brush up before tests, find answers fast, learn key formulas and geometric figures, study quickly, and learn more effectively.

The first part of the book covers the International System of Units (the SI base units, the SI derived units, the SI prefixes, and units outside the SI that are accepted for use with the SI); metric units of measurement; U.S. units of measurements; and tables of equivalent metric and United States Customary System (USCS) units.

The second part of the book covers formulas, rules, and figures related to Algebra, Geometry, Trigonometry, Analytical Geometry, Mathematics of Finance, Calculus, and Statistics.

The third part of the book covers formulas, definitions, and figures related to Mechanics, Fluid Mechanics, Temperature and Heat, Electricity and Magnetism, Light, and Waves and Sound.

Students and professionals alike will find this book a very effective learning tool and reference.
I am grateful to my son Sasha for valuable contributions in the preparation of this book.

Vukota Boljanovic

PART I

UNITS

Units are labels that are used to distinguish one type of measurable quantity from other types. Length, mass, and time are distinctly different physical quantities, and therefore have different unit names, such as meters, kilograms and seconds. We use several systems of units, including the metric (SI) units, the English (or US customary units), and a number of others, which are of mainly historical interest.

This part of the book contains the following:

1. International System of Units
2. Metric Units of Measurement
3. U.S. Units of Measurement
4. Tables of Equivalents

INTERNATIONAL SYSTEM OF UNITS
The International System of Units, abbreviated as SI, is the modernized version of the metric system established by international agreement.

1. SI Base Units

Quantity	Name	Symbol
length	meter	m
mass	kilogram	kg
time	second	s
electric current	ampere	A
thermodynamic temperature	kelvin	K
amount of a substance	mole	mol

2. SI Derived Units

Quantity	Name	Symbol
area	square meter	m^2
volume	cubic meter	m^3
speed, velocity	meter per second	m/s
acceleration	meter per second squared	m/s^2
wave number	reciprocal meter	m^{-1}
mass density	kilogram per cubic meter	kg/m^3

Continued from # 2

specific volume	cubic meter per kilogram	m^3/kg
current density	ampere per square meter	A/m^2
magnetic field strength	ampere per meter	A/m
amount-of-substance concentration	mol per cubic meter	mol/m^3
luminance	candela per square meter	cd/m^2
mass fraction	kilogram per kilogram	kg/kg

3. SI Derived Units with Special Names and Symbols

Quantity	Name	Symbol
plane angle	radian	rad
solid angle	steradian	sr
frequency	hertz	Hz
force	newton	N
pressure, stress	pascal	Pa
energy, work, quantity of heat	joule	J
power, radiant flux	watt	W

International System of Units

Continued from # 3

electric charge, quantity of electricity	coulomb	C
electric potential difference	volt	V
capacitance	farad	F
electric resistance	ohm	Ω
electric conductance	siemens	S
magnetic flux	weber	Wb
magnetic flux density	tesla	T
inductance	henry	H
Celsius temperature	degree Celsius	^{0}C
luminous flux	lumen	lm
illuminance	lux	lx
activity of a radionuclide	becquerel	Bq
absorbed dose, specific energy, kerma	gray	Gy
dose equivalent	sievert	Sv
catalytic activity	katal	kat

4. SI Derived Units Whose Names and Symbols Include SI Derived Units with Special Names and Symbols

Quantity	Name	Symbol
dynamic viscosity	pascal second	$Pa \cdot s$
moment of force	newton meter	$N \cdot m$
angular velocity	radian per second	rad/s
angular acceleration	radian per second squared	rad/s^2
heat flux density, irradiance	watt per square meter	W/m^2
heat capacity, entropy	joule per kelvin	J/K
specific heat capacity, specific entropy	joule per kilogram kelvin	$J/(kg \cdot K)$
specific energy	joule per kilogram	J/kg
energy density	joule per cubic meter	J/m^3
thermal conductivity	watt per meter kelvin	$W/(m \cdot K)$
electric field strength	volt per meter	V/m
electric charge density	coulomb per cubic meter	C/m^3

Continued from # 4

electric flux density	coulomb per square meter	C/m^2
permittivity	farad per meter	F/m
permeability	henry per meter	H/m
molar energy	joule per mole	J/mol
molar entropy, molar heat capacity	joule per mole kelvin	$J/(mol \cdot K)$
exposure (x and γ rays)	coulomb per kilogram	C/kg
absorbed dose rate	gray per second	Gy/s
radiant intensity	watt per steradian	W/sr
radiance	watt per square meter steradian	$W/(m^2 \cdot sr)$

5. Units Outside the SI that Are Accepted for Use with the SI

Name	Symbol	Value in SI units
minute	min	1 min = 60 s
hour	h	1 h = 60 min = 3600 s
day	d	1 d = 24h = 86400 s
liter	L	$1 L = 1 dm^3 = 10^{-3} m^3$
metric tone	t	$1 t = 10^3 kg$
bel	B	1B = 10dB

Continued from # 5

degree (angle)	0	$1^0 = (\pi/180)\,\text{rad}$
minute (angle)	'	$1' = (1/60)^0 =$ $= (\pi/10800)\,\text{rad}$
second (angle)	''	$1'' = (1/60)' =$ $= (\pi/648000)\,\text{rad}$
electronvolt	eV	$1\,\text{eV} = 1.60218 \times 10^{-19}\,\text{J}$
unified atomic mass unit	u	$1\,\text{u} = 1.66054 \times 10^{-27}\,\text{kg}$
astronomical unit	ua	$1\,\text{ua} = 1.49598 \times 10^{11}\,\text{m}$
nautical mile		1 nautical mile = 1852 m
knot		1 knot = 1852/3600 m/s
are	a	$1\,\text{a} = 100\,\text{m}^2$
hectare	ha	$1\,\text{ha} = 100\,\text{a} = 10^4\,\text{m}^2$
bar	bar	$1\,\text{bar} = 10^2\,\text{kPa} = 10^5\,\text{Pa}$
angstrom	$\overset{0}{A}$	$1\,\overset{0}{A} = 0.1\,\text{nm} = 10^{-10}\,\text{m}$
curie	Ci	$1\,\text{Ci} = 3.7 \times 10^{10}\,\text{Bq}$
rad	rad	$1\,\text{rad} = 10^{-2}\,\text{Gy}$
rem	rem	$1\,\text{rem} = 10^{-2}\,\text{Sv}$

6. SI Prefixes

Factor	Name	Symb.	Factor	Name	Symb.
10^1	deca	da	10^{-1}	deci	d
10^2	hecto	h	10^{-2}	centi	c
10^3	kilo	k	10^{-3}	milli	m
10^6	mega	M	10^{-6}	micro	μ
10^9	giga	G	10^{-9}	nano	n
10^{12}	tara	T	10^{-12}	pico	p
10^{15}	peta	P	10^{-15}	fetmo	f
10^{18}	exa	E	10^{-18}	atto	A

METRIC UNITS OF MEASUREMENT

The metric system was first proposed in 1791. The French Revolutionary Assembly adopted it in 1795, and the first metric standards (a standard meter bar and kilogram bar) were adopted in 1799.

7. Units of Length

Name	Symbol	Value
millimeter	mm	1 mm = 0.001 m
centimeter	cm	1 cm = 10 mm

Continued from # 7

decimeter	dm	1 dm = 10 cm
meter	m	1 m = 10 dm = 1000 mm
dekameter	dam	1 dam = 10 m
hectometer	hm	1 hm = 10 dam
kilometer	km	1 km = 10 hm =1000 m

8. Units of Area

Name	Symbol	Value
sq. millimeter	mm^2	$1\,mm^2 = 0.000001\,m^2$
sq. centimeter	cm^2	$1\,cm^2 = 100\ mm^2$
sq. decimeter	dm^2	$1\,dm^2 = 100\,cm^2$
sq. meter	m^2	$1\ m^2 = 100\,dm^2$
sq. decameter	dam^2	$1\,dam^2 = 100\ m^2$
sq. hectometer	hm^2	$1\,hm^2 = 100\,dam^2$
sq. kilometer	km^2	$1\,km^2 = 100\,hm^2$

9. Units of Liquid Value

Name	Symbol	Value
milliliter	mL	1 mL = 0.001L
centiliter	cL	1 cL = 10 mL
deciliter	dL	1 dL = 10 cL
liter	L	1 L = 10 dL = 1000 mL

Continued from # 9

dekaliter	daL	1 daL = 10 L
hectoliter	hL	1 hL = 10 daL
kiloliter	kL	1 kL = 10 hL = 1000 L

10. Units of Volume

Name	Symbol	Value
cu. millimeter	mm^3	1 mm^3 = 10^{-9} m^3
cu. centimeter	cm^3	1 cm^3 = 1000 mm^3
cu. decimeter	dm^3	1 dm^3 = 1000 cm^3
cu. meter	m^3	1 m^3 = 1000 dm^3

11. Units of Mass

Name	Symbol	Value
milligram	mg	1 mg = 0.001g
centigram	cg	1 cg = 10 mg
decigram	dg	1 dg = 10 cg
gram	g	1 g = 10 dg
dekagram	dag	1 dag = 10 g
hectogram	hg	1 hg = 10 dag
kilogram	kg	1 kg = 10 hg = 1000 g
megagram	Mg	1 Mg = 1000 kg = 1t

U.S. UNITS OF MEASUREMENT

Most of the US system of measurements is the same as that for the UK. The biggest differences to be noted are in the present British gallon and bushel--known, as the "Imperial gallon" and "Imperial bushel" are, respectively, about 20 percent and 3 percent larger than the United States gallon and bushel.

12. Units of Length

Name	Symbol	Value
inch	in	1 in = 0.83333 ft
foot	ft	1 ft = 12 in
yard	yd	1 yd = 3 ft
rod	rd	1 rd = 16.5 ft
furlong	fur	1 fur = 40 rd
U.S. mile	mi	1 mi = 8 fur = 5280 ft
nautical mile	nautical mile	1 nautical mile = 1852 m = 6076.1149 ft (appr.)

13. Units of Area

Name	Symbol	Value
sq. inch	in^2	$1\,in^2 = 0.006444\ ft^2$
sq. foot	ft^2	$1\,ft^2 = 144\ in^2$
sq. yard	yd^2	$1\,yd^2 = 9\ ft^2$

Continued from # 13

sq. rood	rd^2	$1\ rd^2 = 272.25\ ft^2$
acre	acre	1 acre = 160 rd $= 43\ 560\ ft^2$
sq. mile	mi^2	$1\ mi^2 = 640$ acre
township		1 township = $6\ mi^2$

14. Units of Liquid Volume

Name	Symbol	Value
gill	gi	1 gi = 0.25 pt
pint	pt	1 pt = 4 gi
quart	qt	1 qt = 2 pt
gallon	gal	1 gal = 4qt = 8 pt = 32 gi

15. Units of Volume

Name	Symbol	Value
cu. inch	in^3	$1\ in^3 = 0.0005787\ ft^3$
cu. foot	ft^3	$1\ ft^3 = 1728\ in^3$
cu. yard	yd^3	$1\ yd^3 = 27\ ft^3$

16. Apothecaries' Units of Liquid Volume

Name	Symbol	Value
minim	min	1 min = 0.016666 dr
fluid dram	fl dr	1 fl dr = 60 min
fluid ounce	fl oz	1 fl oz = 8 fl dr
pint	pt	1 pt = 16 fl oz
quart	qt	1 qt = 2 pt
gallon	gal	1 gal = 4 qt

17. Units of Dry Volume

Name	Symbol	Value
pint	pt	1 pt = 05 qt
quart	qt	1 qt = 2 pt
peck	pk	1 pk = 8 qt
bushel	bu	1 bu = 4 pk

18. Avoirdupois Units of Mass

Name	Symbol	Value
grain	gr	1 gr = 64.79891 mg
dram	dr	1 dr = 27-11/32 gr
ounce	oz	1 oz = 16 dr

Continued from # 18

pound	lb	1 lb = 16 oz
hundredweight	cwt	1 cwt = 100 lb
ton	ton	1ton = 20 cwt = 2000 lb

19. Apothecaries' Units of Mass

Name	Symbol	Value
grain	gr	1 gr = 64.79891 mg
scruple	s ap	1 s ap = 20 dr
apothecaries' dram	dr ap	1dr ap = 3 s ap
apothecaries' ounce	oz ap	1oz ap = 8 dr ap
apothecaries' pound	lb ap	1lb ap = 12 lb ap

20. Troy Units of Mass

Name	Symbol	Value
grain	gr	1 gr = 64.79891 mg
pennyweight	dwt	1 dwt = 24 gr
ounce troy	oz t	1 oz t = 20 dwt
pound troy	lb t	1 lb t = 12 oz t
pennyweight	dwt	1 dwt = 24 gr

TABLES OF EQUIVALENTS
In tables below, all bold equivalents are exact.

1. Units of Length

Name	Equivalents
1 angstrom (Å) =	**0.1** nm **0.0000001** mm 0.000000004 inch
1 centimeter (cm) =	0.393 7 in
1 chain (ch) =	66 ft
1 decimeter (dm) =	3.937 in
1 dekameter (dam) =	32.808 ft
1 fathom =	**6** ft **1.828 8** m
1 furlong (fur) =	**10** ch **660** ft 201.168 m
1 light-year (ly) =	9460730472580.7 km
1 foot (ft) =	**0.304 8** m **12 in** **30.48** cm

Tables of Equivalents

Continued from # 21

1 inch (in) =	**25.4** mm **2. 54** cm
1 kilometer (km) =	0.621 mi
1 meter (m) =	39.37 in 1.094 yd
1 micrometer (μm) =	0.001 mm
1 mile (mi) =	**5, 280** ft 1.609 km
1 mile (international nautical)=	**1. 852** km 1.151 mi
1 millimeter (mm) =	0.03937 in
1 nanometer (nm) =	**0.001** μm 0.000000039 37 in
1 Point (typography) =	**0. 013837** in 1/72 in 0.351 mm
1 rod (rd) =	**16. 5** ft 5.0292 m
1 yard (yd) =	**0. 9144** m

2. Units of Area

Name	Equivalents
1 acre =	**43, 560** ft^2 4, 046 m^2 0.40467 ha
1 are (a) =	119.599 yd^2 0.025 acre
1 hectare (ha) =	2.471 acre
1 square centimeter (cm^2) =	0.155 in^2
1 square foot (ft^2) =	0.0929030 m^2
1 square inch (in^2) =	**645.16** mm^2
1 square kilometer (km^2) =	247.104 acre 0.386 mi^2
1 square meter (m^2) =	1.196 yd^2 10.764 ft^2
1 square mile (mi^2) =	258.999 ha
1 square millimeter (mm^2) =	0.002 in^2
1 square rod (rd^2) =	25.293 m^2
1 square yard (yd^2) =	0.836 m^2

Tables of Equivalents

---

23. Units of Volume

Name	Equivalents
1 barrel (bbl), liquid* =	31 to 42 gal
1 bushel (bu) (U.S.) =	**2,150. 42** in^2 35.239 L
1 cubic centimeter (cm^3) =	0.061 in^3
1 cubic foot (ft^3) =	7.481 gal 28.316 dm^3
1 cubic inch (in^3) =	0.554 fl oz 16.387 cm^3
1 cubic meter (m^3) =	1.308 yd^3
1 cubic yard (yd^3) =	0.765 m^3
1 cup, measuring =	**8** fl oz 237 mL 0.5 lk pt
1 dekaliter (daL) =	2.642 gal 1.135 pk
1 hectoliter (hL) =	26.418 gal 2.838 bu
1 liter (L) =	1.057 fl qt 61.025 in^3
1 milliliter (mL) =	0.271 fl dr 0.061 in^3

Continued from # 23

1 ounce, fluid (fl oz) =	1.805 in^3 29.573 mL
1 peck (pk) =	8.810 L
1 pint (pt), dry =	33.600 in^3 0.551 L.
1 pint (pt), liquid =	**28.875** in^3 0.473 L
1 quart (qt), dry (U.S) =	67.201 in^3 1.101 L
1 quart (qt), liquid (U.S.) =	**57.75** in^3 0.946 L
1 dram, fluid (fl dr) =	**1/8** fl oz 0.226 in^3 3.697 mL
1 gallon (gal) (U.S.) =	**231** in^3 3.785 L **128** fl oz

* There are a variety of "barrels" established by
  law or usage.

24. Units of Mass

Name	Equivalents
1 carat (c) =	200 mg 3.086 gr

Tables of Equivalents

Continued from # 24

1 dram apothecaries (dr ap) =	**60** gr 3.888 g
1 gamma (y) =	**1** μg
1 grain (gr) =	**64.79891** mg
1 gram (g) =	15.432 gr
1 kilogram (kg) =	2.205 lb
1 ounce, troy (oz t) =	**480** gr 31.103 g
1 pennyweight (dwt) =	1.555 g
1 point =	0.01 carat 0.02 mg **7,000** gr
1 pound, troy (lb t) =	**5,760** gr 373.242 g
1 ton, net =	**2,000** lb 0.893 gross ton
1 ton, gross =	**2,240** lb **1.12** net tons 1.016 t
1 ton, metric (t) =	2,204.623 lb 0.984 gross ton 1.102 net tons

PART II

MATHEMATICS

Mathematics is a branch of science large enough to be distinctly separate from "science" and to be placed in its own category.

This part of the book contains the most frequently used formulas, definitions, and rules relating to the following:

1. Algebra
2. Geometry
3. Trigonometry
4. Analytical Geometry
5. Mathematics of Finance
6. Calculus
7. Statistics

ALGEBRA

The purpose of this collection of algebraic references is to provide a brief, clear and handy guide to the more important, formal rules of algebra and the most commonly used formulas for evaluating quantities, as well as examples of their applications for solving algebraic problems.

This section contains the following:

1. Fundamentals of Algebra
2. Determinants
3. Linear Equations
4. Quadratic Equations
5. Inequalities
6. Sequences and Series
7. Functions and Their Graphs

1. Sets of Real Numbers

The set of all rational numbers combined with the set of all irrational numbers gives us the set of real numbers. The relationships among the various sets of real numbers are shown below.

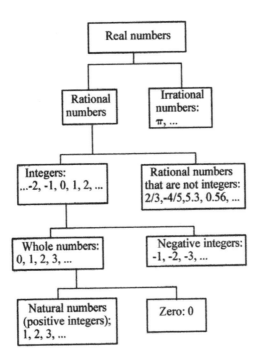

2. Properties of Real Numbers

If a, b, and c are real numbers, then

a) Addition properties

Commutative:	$a + b = b + a$
Associative:	$(a + b) + c = a + (b + c)$
Identity:	$a + 0 = 0 + a = a$
Inverse:	$a + (-a) = (-a) + a = 0$

b) Multiplication properties

Commutative:	$ab = ba$
Associative:	$(ab)c = a(bc)$
Identity:	$a \cdot 1 = 1 \cdot a = a$
Inverse:	$a\left(\dfrac{1}{a}\right) = \left(\dfrac{1}{a}\right)a = 1$
Distributive:	$a(b + c) = ab + ac$

3. Properties of Equality

If a, b, and c are real numbers, then

Identity:	$a = a$
Symmetric:	If $a = b$, then $b = a$
Transitive:	If $a = b$ and $b = c$, then $a = c$
Substitution:	If $a = b$, then a may be replaced by b

Fundamentals of Algebra

4. Properties of Fractions

If $\dfrac{a}{b}$ and $\dfrac{c}{d}$ are fractions of real numbers, where $b \neq 0$ and $d \neq 0$, then

Equality: $\qquad \dfrac{a}{b} = \dfrac{c}{d}$ if and only if $ad = bc$

Equivalence: $\qquad \dfrac{a}{b} = \dfrac{ac}{bc}, \quad (c \neq 0)$

Addition: $\qquad \dfrac{a}{b} + \dfrac{c}{b} = \dfrac{a+c}{b}$

Subtraction: $\qquad \dfrac{a}{b} - \dfrac{c}{b} = \dfrac{a-c}{b}$

Multiplication: $\qquad \dfrac{a}{b} \cdot \dfrac{c}{d} = \dfrac{ac}{bd}.$

Division: $\qquad \dfrac{a}{b} \div \dfrac{c}{d} = \dfrac{a}{b} \cdot \dfrac{d}{c} = \dfrac{ad}{bc}, \quad (c \neq 0)$

Sign:
$$-\dfrac{a}{b} = \dfrac{-a}{b} = \dfrac{a}{-b}$$
$$-\left(\dfrac{-a}{b}\right) = \dfrac{a}{b}$$

5. Division Properties of Zero

If a is a real number, where $a \neq 0$, then

$$\frac{0}{a} = 0$$

(zero divided by any nonzero number is zero).

$$\frac{a}{0} \text{ is undefined}$$

(division by zero is undefined)

$$\frac{0}{0} \text{ is indeterminate.}$$

Relations of this kind, in which there could be any number of values, are called "indeterminate".

6. Real Number Line

The real numbers can be represented by a real number line as shown below.

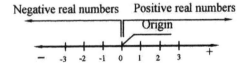

Fundamentals of Algebra

Certain order relationships exist among real numbers.
If a and b are real numbers, then

$$a = b, \qquad \text{if } a - b = 0$$

$$a > b, \qquad \text{if } a - b \text{ is positive}$$

$$a < b, \qquad \text{if } b - a \text{ is positive}$$

The symbols that represent inequality are > (greater than) and < (less than).

7. Interval

In general, there are four interval notations.

a) Open interval
Represents all real numbers between a and b, not including a and not including b. The interval notation is

$$(a, b)$$

b) Closed interval
Represents all real numbers between a and b, including a and including b. The interval notation is

$$[a, b]$$

c) Half-open interval
Represents all real numbers between a and b, not including a but including b. The interval notation is

$$(a, b]$$

d) Half-closed interval
Represents all real numbers between a and b, including a but not including b. The interval notation is

$$[a, b)$$

8. Absolute Value

The absolute value of the real number a, denoted $|a|$, is defined by

$$|a| = \begin{cases} a & \text{if } a \geq 0 \\ -a & \text{if } a < 0 \end{cases}$$

a) Properties of absolute value
For all real numbers a and b,

Product: $\quad |ab| = |a| \cdot |b|$

Quotient: $\quad \left|\dfrac{a}{b}\right| = \dfrac{|a|}{|b|}, \quad (b \neq 0)$

Difference: $\quad |a - b| = |b - a|$

Fundamentals of Algebra

Inequality:
$$|a + b| \leq |a| + |b|$$
$$|-a| = |a|$$

If $|a| = b$, then $a = b$ or $a = -b$

If $|a| < b$, then $-b < a < b$

If $|a| > b$, then $a > b$ or $a < -b$

9. Distance between Two Points on the Number Line

For any real number a and b, the distance between a and b denoted by $d(a, b)$, is

$$d|a, b| = |a - b|, \text{ or equivalently, } |b-a|$$

10. Definition of Positive Integer Exponents

For any positive integer n, and if b is any real number, then,

$$b^n = b \cdot b \cdot b....b, \qquad (n \text{ factors of } b)$$

where

$b =$ the base

$n =$ the exponent

11. Definition of b^0

For any nonzero real number b,

$$b^0 = 1$$

12. Definition of b^{-n}

For any natural number n,

$$b^{-n} = \frac{1}{b^n} \text{ and } \frac{1}{b^{-n}} = b^n, \quad (b \neq 0)$$

Note: The expressions 0^0, 0^n where n is a negative integer, and $\dfrac{x}{0}$ are all undefined expressions.

13. Properties of Exponents

If m, n, and p are integers and a and b are real numbers, then,

Product:
$$a^m a^n = a^{m+n}$$
$$a^m \cdot b^m = (ab)^m$$

Quotient:
$$\frac{a^m}{a^n} = a^{m-n}, \quad (a \neq 0)$$

Fundamentals of Algebra

$$\frac{a^m}{b^m} = \left(\frac{a}{b}\right)^m, \quad (b \neq 0)$$

Power:
$$\left(a^m\right)^n = a^{mn}$$

$$\left(a^m b^n\right)^p = a^{mp} b^{np}$$

$$\left(\frac{a^m}{b^n}\right)^p = \frac{a^{mp}}{b^{np}}, \quad (b \neq 0)$$

14. Definition of $\sqrt[n]{a}$

The symbol $\sqrt[n]{a}$ is called a radical. $\sqrt{}$ is the radical sign, n is the index or root (which is omitted when it is 2), and a is the radicand.

$$\sqrt[n]{a} = a^{\frac{1}{n}}$$

15. Properties of Radicals

If m and n are natural numbers greater than or equal to 2, and a and b are nonnegative real numbers, then

Product:
$$\sqrt[n]{a} \cdot \sqrt[n]{b} = \sqrt[n]{ab}$$

Quotient: $$\frac{\sqrt[n]{a}}{\sqrt[n]{b}} = \sqrt[n]{\frac{a}{b}} = \left(\frac{a}{b}\right)^{\frac{1}{n}}, \quad (b \neq 0)$$

Index: $$\sqrt[n]{\sqrt[m]{a}} = \sqrt[n \cdot m]{a}$$

$$\left(\sqrt[n]{a}\right)^m = a^{\frac{m}{n}}$$

$$\sqrt[n]{a^n} = a$$

$$\left(\sqrt[n]{a}\right)^n = a$$

16. General Form of a Polynomial

The general form of a polynomial of degree n in the variable x is

$$a_n x^n + a_{n-1} x^{n-1} + \ldots + a_1 x + a_0$$

Note:

n is a nonnegative integer and $a_n \neq 0$.

The coefficient a_n is the leading coefficient, and

a_0 is the constant term.

17. Factoring Polynomials

Factoring a polynomial is writing a polynomial as a product of polynomials of lower degree.

---

a) The square of a binomial:

$$(a \pm b)^2 = a^2 \pm 2ab + b^2$$

b) The cube of a binomial:

$$(a \pm b)^3 = a^3 \pm 3a^2b + 3ab^2 \pm b^3$$

c) The difference of two squares:

$$a^2 - b^2 = (a + b)(a - b)$$

d) The sum or difference of two cubes:

$$a^3 \pm b^3 = (a \pm b)(a^2 \mp ab + b^2)$$

e) The square of a trinomial:

$$(a \pm b + c)^2 = a^2 \pm 2ab + 2ac + b^2 \pm 2bc + c^2$$

18. Order of Operations

If grouping symbols are present, evaluate by performing the operations within the grouping symbol first, while observing the order given in Steps 1 to 3. For example,

$$2x^3 - \left\{x^2 - \left[x - (2x - 1)\right] + 4\right\}$$

Step 1: Remove parenthesis

$$= 2x^3 - \left\{x^2 - \left[x - 2x + 1\right] + 4\right\}$$
$$= 2x^3 - \left\{x^2 - \left[-x + 1\right] + 4\right\}$$

Step 2: Remove brackets

$$= 2x^3 - \left\{x^2 + x - 1 + 4\right\}$$
$$= 2x^3 - \left\{x^2 + x + 3\right\}$$

Step 3: Remove braces

$$= 2x^3 - x^2 - x - 3$$

The operations of multiplication and division take precedence over addition and subtraction.

Fundamentals of Algebra

19. Adding and Subtracting Polynomials

Adding:
$$\left(ax^2 - bx - c\right) + \left(a_1 x^2 - b_1 x - c\right) =$$
$$= \left(a + a_1\right)x^2 + \left(b - b_1\right)x - 2c$$

Subtracting:
$$\left(ax^2 + bx - c\right) - \left(a_1 x^2 - b_1 x - c\right) =$$
$$= \left(a - a_1\right)x^2 + \left(b + b_1\right)x$$

20. Multiplying Polynomials

$$(ax^2 - bx + c) \cdot \left(a_1 x^2 - 1\right) =$$
$$ax^2(a_1 x^2 - 1) - bx(a_1 x^2 - 1) + c(a_1 x^2 - 1) =$$
$$= aa_1 x^4 - a_1 bx^3 - (a - a_1 c)x^2 + bx - c$$

21. Dividing Polynomials

For examples:

a) Let polynomial $\left(x^2 - 9x + 10\right)$ be divided by polynomial $\left(x + 1\right)$, and

b) Let polynomial $\left(ax^5 + bx^3 - c\right)$ be divided by monomial $a_1 x$, then

a) Dividing a polynomial by a polynomial:

$$(x^2 - 9x + 10) \div (x + 1) = x - 10$$

$x^2 + x$

\- \- (changed sign)

$-10x - 10$
$-10x - 10$

$+$ $+$ (changed sign)

0

b) Dividing a polynomial by a monomial:

$$(ax^5 + bx^3 - c) \div a_1 X = \frac{ax^5}{a_1 X} + \frac{bx^3}{a_1 X} - \frac{c}{a_1 X} =$$

$$\frac{a}{a_1} x^4 + \frac{b}{a_1} x^2 - \frac{c}{a_1 X}$$

22. Rational Expressions

A rational expression is a fraction in which the numerator and denominator are polynomials.

For example:

$$\frac{x^2 - 4x - 21}{x^2 - 9}, \text{ or } \frac{p}{q}$$

Fundamentals of Algebra

---

a) Properties of rational expressions

Let $\dfrac{p}{q}$ and $\dfrac{r}{s}$ be rational expressions where

$q \neq 0$ and $s \neq 0$

Equality: $\qquad\qquad\qquad \dfrac{p}{q} = \dfrac{r}{s}$ if and only if $ps = qr$

Equivalent expressions: $\qquad \dfrac{p}{q} = \dfrac{pr}{qr}, \; r \neq 0$

Sign : $\qquad\qquad\qquad\qquad -\dfrac{p}{q} = \dfrac{-p}{q} = \dfrac{p}{-q}$

b) Operation with rational expressions

For all rational expressions $\dfrac{p}{q}$ and $\dfrac{r}{s}$, where

$q \neq 0$ and $s \neq 0$

Addition: $\qquad\qquad\qquad \dfrac{p}{q} + \dfrac{r}{q} = \dfrac{p+r}{q}$

Subtraction: $\qquad\qquad\quad \dfrac{p}{q} - \dfrac{r}{q} = \dfrac{p-r}{q}$

Multiplication:
$$\frac{p}{q} \cdot \frac{r}{s} = \frac{pr}{qs}$$

Division
$$\frac{p}{q} \div \frac{r}{s} = \frac{ps}{qr}, \qquad r \neq 0$$

c) Least common denominator (LCD)

Adding and subtracting rational expressions when denominators are differ; we must find equivalent rational expressions that have a common denominator. It is most efficient to find the LCD of the expressions:

Step 1: Factor each denominator completely and express repeated factors using exponential notation.

Step 2: Identify the largest power of each, factoring any single factorization. The LCD is the product of each factor raised to the largest power.

Example:

Find LCD and add rational expressions:
$$\frac{3}{x^2 + x} \text{ and } \frac{2}{x^2 - 1}$$

Solution:

Step 1:
$$x^2 + 1 = x(x+1), \text{ and}$$
$$x^2 - 1 = x(x-1)$$

Step 2: The LCD of the two expressions is

$$x(x+1)(x-1)$$

For adding fractions, we express each fraction using the common denominator, and then we add the numerators.

$$\frac{3}{x^2+x} + \frac{2}{x^2-1} = \frac{3}{x(x+1)} + \frac{2}{(x+1)(x-1)}$$

$$= \frac{3(x-1)+2x}{x(x+1)(x-1)} = \frac{5x-3}{x(x^2-1)}$$

23. Complex Fractions

A complex fraction is a fraction whose numerator or denominator or both contain more fractions.

To simplify a complex fractions use one of two methods:

Method 1: Find the LCD of all the denominators within the complex fraction. Then multiply both the numerator and denominator of the complex fraction by the LCD.

Method 2: First add or subtract, if necessary, to get a single fraction in both the numerator and the denominator. Then divide by multiplying by the reciprocal of the denominator.

Example: Simplify a complex fraction $\dfrac{3 - \dfrac{1}{a}}{1 + \dfrac{4}{a}}$

Solution:

$$\frac{3 - \dfrac{1}{a}}{1 + \dfrac{4}{a}} = \frac{\dfrac{3a-1}{a}}{\dfrac{a+4}{a}} = \frac{(3a-1)a}{(a+4)a} = \frac{3a-1}{a+4}$$

24. Definition of a Complex Number

A complex number is any number that can be written

$$z = a + bi$$

where

  a = real part of the complex number

  b = real number of imaginary part of the complex number

  i = imaginary unit $\left(i = \sqrt{-1}\right)$

  a) Operations with complex numbers

Let $a + bi$ and $c + di$ be complex numbers, then,

Addition: $(a + bi) + (c + di) = (a + c) + (b + d) \cdot i$

Subtraction: $(a + bi) - (c + di) = (a - c) + (b - d) \cdot i$

Multiplication: $(a+bi) \cdot (c+di) = (ac-bd)+(ad+bc) \cdot i$

Division: $\dfrac{a+bi}{c+di} = \dfrac{ac+bd}{c^2+d^2} + \dfrac{bc-ad}{c^2+d^2} i, \ (c+di \neq 0)$

b) Conjugate of a complex number

The conjugate of a complex number $z = a+bi$ is

$$\bar{z} = a-bi$$

Properties: $z + \bar{z}$ is a real number

$z \cdot \bar{z} = |z|^2$ is always real number

$\bar{z} = z$ if and only if z is a real number

$\bar{z}^n = (\bar{z})^n$ for all natural numbers

c) Powers of i

If n is a positive integer, then,

$$i^n = i^r$$

where

    r = remainder of the division of n by 4

Example: Evaluate i^{37}
Use the theorem on powers of i
$i^{37} = i^1 = i$ (the remainder of $37 \div 4$ is 1)

25. Definition of a Linear Equation

An equation is a statement of equality between two mathematical expressions.

A linear equation in the single variable x can be written in the form

$$ax + b = 0$$

where

a, b = real numbers $(a \neq 0)$

26. Addition and Multiplication Properties of Equality

If $\quad a = b$, then $\qquad a + c = b + c$

If $\quad a = b$, then $\qquad ac = bc$

If $\quad -a = b$, then $\qquad a = -b$

If $\quad x + a = b$, then $\quad x = b - a$

If $\quad x - a = b$, then $\quad x = a + b$

If $\quad ax = b$, then $\qquad x = \dfrac{b}{a}$

If $\dfrac{x}{a} = b$, then $\qquad x = ab$

27. Systems of Linear Equations

A system of linear equations can be solved in various different ways, such as by substitution, elimination, determinants, matrices, graphing, etc.

a) The method of substitution:

$$x + 2y = 4 \qquad (1)$$
$$3x - 2y = 4 \qquad (2)$$

The method of substitution involves five steps:

Step 1: Solve for y in equation (1)

$$y = \frac{4 - x}{2}$$

Step 2: Substitute this value for y in equation (2). This will change equation (2) to an equation with just one variable, x

$$3x - 2\frac{4 - x}{2} = 4$$

Step 3: Solve for x in the translated equation (2)

$$4x = 8$$

$$x = 2$$

Step 4: Substitute this value of x in the y equation obtained in Step 1

$$2 + 2y = 4$$

$$y = 1$$

Step 5: Check answers by substituting the values of x and y in each of the original equations. If, after the substitution, the left side of the equation equals the right side of the equation, the answers are correct.

b) The method of elimination:

$$x + 2y = 4 \qquad (1)$$
$$3x - 2y = 4 \qquad (2)$$

The process of elimination involves four steps:

Step 1: Change equation (1) by multiplying it by (-3) to obtain a new and equivalent equation (1).

$$-3x - 6y = -12, \quad \text{new equation (1).}$$

Step 2: Add new equation (1) to equation (2) to obtain equation (3).

$$-3x - 6y = -12$$
$$3x - 2y = 4$$

$$-8y = -8 \qquad (3)$$
$$y = 1$$

Step 3: Substitute $y = 1$ in equation (1) and solve for x.

$$x + 2 \cdot 1 = 4$$
$$x = 2$$

Step 4: Check your answers in equation (2).

$$3 \cdot 2 - 2 \cdot 1 = 4$$
$$4 = 4$$

28. Determinants

Let system (1) be

$$a_{11}x + a_{12}y = r_1$$
$$a_{21}x + a_{22}y = r_2 \qquad (1)$$

and represent any system of linear equations, then the second order determinant of system (1) is

ALGEBRA
Determinants

$$D = \begin{vmatrix} a_{11} & a_{12} \\ a_{21} & a_{22} \end{vmatrix} = a_{11} \cdot a_{22} - a_{21} \cdot a_{12}$$

$$\underset{-}{} \qquad \underset{+}{}$$

To solve for x, insert column r in place of column x into determinant D then

$$D_x = \begin{vmatrix} r_1 & a_{12} \\ r_2 & a_{22} \end{vmatrix} = r_1 \cdot a_{22} - r_2 \cdot a_{12}$$

$$\underset{-}{} \qquad \underset{+}{}$$

$$x = \frac{D_x}{D}, \quad (D \neq 0)$$

To solve for y, insert column r in place of column y into determinant D, then

$$D_y = \begin{vmatrix} a_{11} & r_1 \\ a_{21} & r_2 \end{vmatrix} = a_{11} \cdot r_2 - a_{21} \cdot r_1$$

$$\underset{-}{} \qquad \underset{+}{}$$

$$y = \frac{D_y}{D}, \quad (D \neq 0)$$

Example:

Solve system equations by determinants:

$$2x + 4y = 8$$
$$3x - 2y = 4$$

Solution:

Determinant for system equations is

$$D = \begin{vmatrix} 2 & 4 \\ 3 & (-2) \end{vmatrix} = 2 \cdot (-2) - 3 \cdot 4 = -16$$

Determinant for x is

$$D_x = \begin{vmatrix} 8 & 4 \\ 4 & (-2) \end{vmatrix} = 8 \cdot (-2) - 4 \cdot 4 = -32$$

$$x = \frac{D_x}{D} = \frac{-32}{-16} = 2$$

Determinant for y is

$$D_y = \begin{vmatrix} 2 & 8 \\ 3 & 4 \end{vmatrix} = 2 \cdot 4 - 3 \cdot 8 = -16$$

$$y = \frac{D_y}{D} = \frac{-16}{-16} = 1$$

29. Quadratic Equations

The standard form of quadratic equations is

$$ax^2 + bx + c = 0$$

where

a, b, c = constants ($a \neq 0$)

a) Solving quadratic equations by factoring.

Let $x^2 - 3x + 2 = 0$ be the standard form of a quadratic equation, then,

$$x^2 - 3x + 2 = x^2 - 2x - x + 2 = 0$$
$$(x - 2)(x - 1) = 0$$

The roots of the equation are:

$$(x - 2) = 0$$
$$x = 2,$$

and

$$(x - 1) = 0$$
$$x = 1$$

b) Solving quadratic equations using Vieta's rule.
Normal form of quadratic equation:

$$x^2 + px + q = 0$$

Solutions:

$$x_{1,2} = -\frac{p}{2} \pm \sqrt{\frac{p^2}{4} - q}$$

Quadratic Equations

Vieta's rule:

$$p = -\left(x_1 + x_2\right)$$
$$q = x_1 \cdot x_2$$

c) Solving quadratic equations by completing the square.
Let the standard form of quadratic equations be

$$ax^2 + bx + c = 0$$

Step 1: Write the equation in the form

$$x^2 + \frac{b}{a}x = \frac{c}{a}$$

Step 2: Square half of the coefficient of x.

Step 3: Add the number obtained in step 2 to both sides
of the equation, factor, and solve for x.

Example:
Solve the quadratic equation by completing the square:

$$x^2 - 2x - 2 = 0$$

Solution:

Step 1: $\qquad\qquad x^2 - 2x = 2$

Step 2:
$$\left(-\frac{2}{2}\right)^2 = 1$$

Step 3:
$$x^2 - 2x + 1 = 2 + 1$$
$$(x-1)^2 = 3$$
$$x_{1,2} = 1 \pm \sqrt{3}$$
$$x_1 = 1 + \sqrt{3}$$
$$x_2 = 1 - \sqrt{3}$$

d) Solving quadratic equations by using the quadratic formula.
The quadratic equation

$$ax^2 + bx + c = 0$$

with real coefficients and $a \neq 0$, can be solved as follows:

$$x_{1,2} = \frac{-b \pm \sqrt{b^2 - 4ac}}{2a}$$

where

$b^2 - 4ac$ = discriminant D of the quadratic equation.

Quadratic Equations

If $D = b^2 - 4ac > 0$, then the quadratic equation has two real and distinct roots.

If $D = b^2 - 4ac = 0$, then the quadratic equation has a real root that is a double root.

If $D = b^2 - 4ac < 0$, then the quadratic equation has two distinct but no real roots.

Example:
Classify the roots of each quadratic equation:

$$1)\ 2x^2 - 5x + 1 = 0$$
$$2)\ 3x^2 + 6x + 7 = 0$$

Solution:

1) $D = b^2 - 4ac = (-5)^2 - 4(2)(1) = 25 - 8 = 17$

$D = 17 > 0$

because $D > 0$, quadratic equation

$2x^2 - 5x + 1 = 0$ has two distinct real roots.

2) $D = b^2 - 4ac = (6)^2 - 4(3)(7) = 36 - 84 = -48$

$D = -48 < 0$

because $D < 0$, quadratic equation

$3x^2 + 6x + 7 = 0$, has two distinct but no real roots.

---

30. Properties of Inequalities

For real numbers a, b, and c, the properties of inequalities follow:

If $a < b$, then $\quad a + c < b + c$
   (Adding the same number to each side of an
     inequality preserves the order of the inequality.)

If $a < b$, and if $c > 0$, then $\quad ac < bc$

   (Multiplying each side of an inequality by the same
     positive number preserves the order of the
     inequality.)

If $a < b$ and $b < c$, then $\quad a < c$

If $a < b$ and $c < d$, then $\quad a + c < b + d$

If $0 < a < b$ and $0 < c < d$, then $\quad ac < bd$

31. Arithmetic Sequence

The sequence 1, 4, 7, 10, …is an example of an arithmetic sequence or arithmetic progression. The difference between the successive terms is the same, constant d. In general, an arithmetic sequence is

$$a_1, (a_1 + d), (a_1 + 2d), (a_1 + 3d), \ldots$$

Sequence and Series

The nth term of an arithmetic sequence is

$$a_n = a_1 + (n-1)d$$

where

$d =$ common difference $[d = (a_n - a_{n-1})]$

$a_1 =$ the first term

a) Arithmetic mean

Each term of an arithmetic sequence is the arithmetic mean of its adjacent terms:

$$a_m = \frac{a_{m-1} + a_{m+1}}{2}$$

where

$a_m =$ arithmetic mean

$a_{m-1}, a_{m+1} =$ adjacent terms.

32. Arithmetic Series

The sum of the arithmetic sequence is called an arithmetic series.

a) Sum of the first n terms

The sum of the terms of an arithmetic sequence is given by the formula:

$$S_n = \frac{n}{2}(a_1 + a_n)$$

where

$$a_n = a_1 + (n-1)d, \quad (n = 1, 2, 3, ..)$$

An alternative formula for the sum of an arithmetic series is

$$S_n = \frac{n[2a_1 + (n-1)d]}{2}$$

Example:

Find S_{20} for the arithmetic sequence whose first term is $a_1 = 3$ and whose common difference is $d = 5$.

Solution:

Substituting $a_1 = 3$, $d = 5$, and $n = 20$ in the formula,

$$S_{20} = \frac{20}{2}[2(3) + (20-1)5] = 1010$$

33. Geometric Sequences

The sequence: a_1, $a_1 r$, $a_1 r^2$, $a_1 r^3$,..., $a_1(r^{n-1})$,... is called a geometric sequence. The ratio between two successive terms is the same constant. This constant is called the common ratio.

a) The nth term of a geometric sequence is

$$a_n = a_1 r^{n-1}$$

where

$a_1 = $ the first term

$r = $ common ratio ($r = \dfrac{a_{i+1}}{a_i}$)

b) Geometric mean

Each term of a geometric sequence is the geometric mean of its adjacent terms:

$$a_m = \sqrt{a_{m-1} \cdot a_{m+1}} , \quad (1 < m < n)$$

where

$a_m = $ geometric mean

$a_{m-1}, a_{m+1} = $ adjacent terms

34. Geometric Series

The sum of geometric sequences is called a geometric series.

$$S_n = a_1 + a_1 r + a_1 r^2 + ... + a_1 r^{n-2} + a_1 r^{n-1}$$

a) Sum of n terms:

$$S_n = a_1 \frac{1 - r^n}{1 - r}, \quad (r \neq 1)$$

where

$$r = \text{the common ratio } (r = \frac{a_{i+1}}{a_i})$$

Example:

Bob saves \$150 in January and each month thereafter, Bob manages to save half of what he saved the previous month. How much does Bob save in the 12th month, and what is his total savings after 12 months?

Solution:

The amounts saved each month form a geometric sequence with $a_1 = 150$, $r = 0.5$ and $n = 12$: then

$$a_n = a_1 r^{n-1} =$$

$$a_{12} = 150\left(\frac{1}{2}\right)^{12-1} = 150\left(\frac{1}{2}\right)^{11} = 0.073$$

This means that Bob saves 0.073 cents in the 12th month. Bob's total savings is:

$$S_{12} = 150\frac{1-\left(\frac{1}{2}\right)^{12}}{1-\frac{1}{2}} = 299.9267$$

The total amount saved is \$299.93

---

b) Sum of an infinite geometric series

If a_n is a geometric sequence with $|r| < 1$, $n \to \infty$ and first term a_1, then the sum of the infinite geometric series is

$$S = \frac{a_1}{1-r}$$

35. Binomial Theorem

For any binomial $a + b$ and any natural number n,

$$(a+b)^n = \binom{n}{0}a^n b^0 + \binom{n}{1}a^{n-1}b^1 + \binom{n}{2}a^{n-2}b^2 + \ldots$$

$$+ \binom{n}{n-1}a^1 b^{n-1} + \binom{n}{n}a^0 b^n$$

$$= \sum_{k=0}^{n} \binom{n}{k}a^{n-k}b^k$$

where

$$k = \text{binomial coefficient.} \quad \binom{n}{k} = \frac{n!}{k!(n-k)!}.$$

a) A specific term of a binomial expansion

The $(k+1)$st term of the expansion of $(a+b)^n$ is given by

$$\binom{n}{k} a^{n-k} b^k$$

Example:

Find the fifth term in the expansion of $\left(2x^3 - 5y^2\right)^6$

Solution:

First, we note that 5 = 4 + 1. Thus,

$a = 2x^3$, $b = -3y^2$, $k = 4$, and $n = 6$ we have

$$\binom{n}{k} a^{n-k} b^k =$$

$$\binom{6}{4}\left(2x^3\right)^2\left(-3y\right)^4 =$$

$$\frac{6!}{4!(6-4)!}\left(2x^3\right)^2\left(-3y^2\right)^4 =$$

$$15\left(4x^6\right)\left(81y^8\right) = 4860x^6y^8$$

The fifth term is $4860x^6y^8$

36. The Cartesian Coordinate System

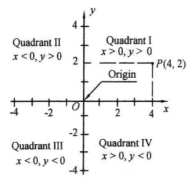

The Cartesian coordinate system in two dimensions, also known as a rectangular coordinate system, is commonly defined by two axes at right angles to each other and forming an xy-plane. The horizontal axis is labeled x, and the vertical axis is labeled y. The point of intersection, where the axes meet, is called the *origin* and is normally labeled 0.

To plot a point $P(a, b)$ means to draw a dot at its location in the coordinate plan. In the figure we have plotted the point $P(4, 2)$.

37. Linear Functions

A linear function is a function that can be represented by a linear equation of the form

$$f(x) = y = mx + b$$

where
m and b = real constants.

The graph of function $f(x) = y = mx + b$ is

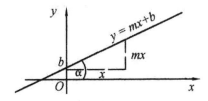

38. Forms of Linear Equations

General form: $Ax + By + C = 0$

where
A, B, C = constants, ($A \neq 0$, $B \neq 0$)

Slope intercept form is

$$y = mx + b$$

where
m = slope of the line, ($m = \tan \alpha$)
b = intercept on the y-axis

Vertical line: $x = a$

Horizontal line: $y = b$

Point-slope form: $y - y_1 = m(x - x_1)$

Intercept form: $\dfrac{x}{a} + \dfrac{y}{b} = 1$

Two-point form: $y - y_1 = \left(\dfrac{y_2 - y_1}{x_2 - x_1} \right)(x - x_1)$

39. Quadratic Functions

A quadratic function is a non-linear function that can be represented by an equation of the form

$$f(x) = y = ax^2 + bx + c, \qquad a \neq 0$$

where

    $a, b, c =$ real numbers

The graph of $f(x) = y = ax^2 + bx + c$ function is a parabola.

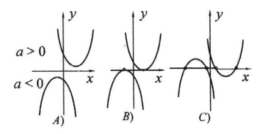

a) Properties of the quadratic functions:

1. If $a > 0$, the parabola opens upward
2. If $a < 0$, the parabola opens downward
3. The vertex of the parabola is $\left[-\dfrac{b}{2a}, f\left(-\dfrac{b}{2a}\right)\right]$
4. The axis of symmetry of the parabola is $x = -\dfrac{b}{2a}$

5. The x--intercepts are found by solving $f(x) = 0$
6. The y--intercept is $f(0) = c$

b) Using the discriminant

When a, b, and c in equation $ax^2 + bx + c = 0$ are real numbers, then a graph of $f(x) = y = ax^2 + bx + c$ can be done in three ways:

If $b^2 - 4ac < 0$, the graph of $f(x)$ does not cross the
x-axis (Fig. A)

If $b^2 - 4ac = 0$, the graph of $f(x)$ touches the x-axis at
one point (Fig. B)

If $b^2 - 4ac > 0$, the graph of $f(x)$ crosses the x-axis at
two points (Fig. C)

40. Basic Operation of Functions

If the ranges of functions f and g are subsets of the real
numbers, then

Sum: $(f + g)(x) = f(x) + g(x)$

Difference: $(f - g)(x) = f(x) - g(x)$

Product: $(f \cdot g)(x) = f(x) \cdot g(x)$

Quotient: $\left(\dfrac{f}{g}\right)(x) = \dfrac{f(x)}{g(x)}, \quad g(x) \neq 0$

41. Exponential Functions

The function defined by

$$f(x) = y = a^x, \quad (a \neq 1)$$

is called an exponential function.

where
  a = base (positive constant)
  x = exponent (any real number)

The graph of $f(x) = y = a^x$, $(a \neq 1)$ is

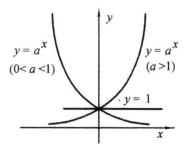

  a) Properties of the exponential functions
Exponential function $f(x) = y = a^x$, $(a \neq 1)$ has the
following properties:
1. The domain consists of all real numbers x, $(-\infty, \infty)$.
2. The range consists of all positive numbers $(0, \infty)$.
3. The function increases when $a > 1$, and it decreases
   when $0 < a < 1$.
4. The graph passes through point $(0, 1)$.

42. Natural Exponential Function
The function defined by

$$f(x) = e^x$$

is called the natural exponential function,

where
e = base (e = 2.71828183…)
x = exponent (any real number).

43. Logarithmic Functions

The function defined by

$$f(x) = y = \log_a x \text{ if and only if } a^y = x$$

is called the logarithmic function,

where
a = base
x = argument (any real number)

A graph of $y = \log_a x$ is

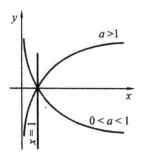

a) Properties of the logarithmic functions

Logarithmic function $y = \log_a x$, $\left(a \neq 1 \right)$ has the following properties:

1. The domain consists of all positive numbers $x, (0, \infty)$.
2. The range consists of all real numbers y $(-\infty, \infty)$.
3. The function increases from left to right if $a > 1$, and it decreases from left to right if $0 < a < 1$.
4. The graph passes through point $(1, 0)$.
5. The graph is an unbroken curve devoid of holes or breaks.

b) Laws of logarithms:

$$\log_a (x \cdot y) = \log_a x + \log_a y$$

$$\log_a \frac{x}{y} = \log_a x - \log_a y$$

$$\log_a x^n = n\log_a x$$

$$\log_a(x \cdot y) = \log_a x + \log_a y$$

$$\log_a \frac{x}{y} = \log_a x - \log_a y$$

$$\log_a x^n = n\log_a x$$

$$\log_a \sqrt[n]{x} = \frac{1}{n}\log_a x$$

$$\log_a 1 = 0$$

$$\log_a a = 1$$

c) Logarithmic Notation:

$\log x = \log_{10} x$ (common logarithm)

$\ln x \equiv \log_e x$ (natural logarithm)

GEOMETRY

Geometry is the branch of mathematics concerned with the properties of and relationships between points, lines, planes, angles, and solids and with generalizations of these concepts.

If geometry has always been your nemesis, here we will explain simply and easily how to do every kind of geometrical problem you are likely to face in the performance of your professional job or study of mathematics in a high school or college, from angles to solid bodies.

This section contains the most frequently used formulas, rules, and definitions regarding to the following:

1. Angles
2. Areas
3. Solid Bodies

Angles

---

1. Definition of an Angle

Two rays that share the same endpoint form an angle. The point where the rays intersect is called the *vertex* of the angle. The two rays are called the *sides* of the angle.

2. Unit Measurement of Angles

The radian measure of the angle φ is the ratio of the arc length to the radius.

$$\varphi_{(\text{radian})} = 2\pi \cdot \frac{\phi^0}{360^0}$$

$$1 \text{ radian} = 57.2957^0$$

$$\phi^0 = 360^0 \cdot \frac{\varphi_{(\text{radian})}}{2\pi}$$

3. Acute Angles

An acute angle is an angle measuring between 0 and 90 degrees.

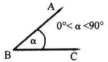

4. Obtuse Angles

An obtuse angle is an angle measuring between 90 and 180 degrees.

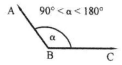

5. Right Angles

A right angle is an angle measuring exactly 90 degrees.

6. Complementary Angles

Two angles are called complementary angles if the sum of their degree measurements equals 90 degrees.

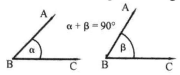

7. Supplementary Angles

Two angles are called supplementary angles if the sum of their degree measurements equals 180 degrees.

Angles

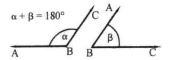

8. Vertical Angles

For any two lines that meet, such as in the diagram below, angle α and angle β are called vertical angles. Vertical angles have the same degree measurement. Angle γ and angle δ are also vertical angles.

9. Alternate Interior Angles

For any pair of parallel lines 1 and 2 that are both intersected by a third line, such as line 3 in the diagram below, angle α and angle β are called alternate interior angles. Alternate interior angles have the same degree measurement. Angle γ and angle δ are also alternate interior angles.

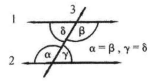

---

10. Alternate Exterior Angles

For any pair of parallel lines 1 and 2 that are both intersected by a third line, such as line 3 in the diagram below, angle α and angle β are called alternate exterior angles. Angle γ and angle δ are also alternate exterior angles.

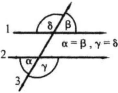

11. Corresponding Angles

For any pair of parallel lines 1 and 2 that are both intersected by a third line, such as line 3 in the diagram below, angle α and angle β are called corresponding angles. Angle γ and angle δ are also corresponding angles.

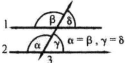

12. Angle Bisector

An angle bisector is a ray that divides an angle into two equal angles.

Areas

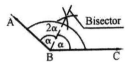

13. Perpendicular Angles

Two angles whose rays meet at a right angle are
perpendicular.

14. Thales' Theorem

A triangle inscribed in a semicircle with a radius R, and
diameter d is a right triangle, as is shown below:

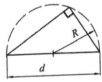

15. Oblique Triangle

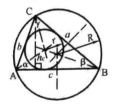

An oblique triangle is any triangle that is not a right triangle. It could be an acute triangle (all three angles of the triangle are smaller than right angles) or it could be an obtuse triangle (one of the three angles is greater than a right angle).

a) Circumscribed circle

The point where perpendicular bisectors to each side of a triangle meet is the center of a circle that circumscribes a triangle. The radius R of a circumscribed circle around a triangle is

$$R = \frac{abc}{4A}$$

where
$$a, b, c, = \text{sides of a triangle}$$
$$A = \text{surface of a triangle}$$

b) Inscribed circle

The point where bisectors of 3 angles of a triangle meet is the center of an inscribed circle in the triangle.
A radius r of a inscribed circle in a triangle is

$$r = \frac{A}{s}$$

where

$$s = \frac{a+b+c}{2}$$

---

c) Sum of the angles in a triangle

$$\alpha + \beta + \delta = 180^{0}$$

16. Geocenter of a Triangle

The medians of a triangle are lines from each vertex to the midpoint of the opposite side. The medians always intersect in a single point called the centroid, or geocenter.

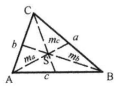

17. Orthocenter

The three altitudes intersect in a single point called the orthocenter of the triangle.

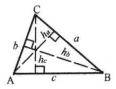

18. Similarity of Triangles

Two triangles are said to be similar:

1. If and only if the angles of one are equal to the corresponding angles of the other. In this case, the lengths of their corresponding sides are proportional.

---

2. When two triangles share an angle and the sides opposite to that angle are parallel.

3. If two angles in two different triangles are the same: in that case then the triangles are similar, too.

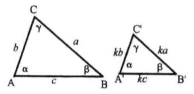

19. The Law of Cosines

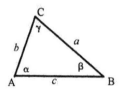

The law of cosines is valid for all triangles, even if any angle of the triangle is not a right angle.

The law of cosines can be used to compute the side lengths and angles of a triangle if all three sides or two sides and an enclosed angle are known.

$$a^2 = b^2 + c^2 - 2bc\cos\alpha$$
$$b^2 = a^2 + c^2 - 2ac\cos\beta$$
$$c^2 = a^2 + b^2 - 2ab\cos\gamma$$

20. The Law of Sines

The law of sines can be used to compute the side lengths for a triangle as two angles and one side are known. If two sides and an unenclosed angle are known, the law of sines may also be used.

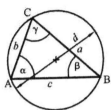

$$\frac{\sin \alpha}{a} = \frac{\sin \beta}{b} = \frac{\sin \gamma}{c} = \frac{1}{d}$$

where

a, b, c = sides of the triangle

d = the diameter of the circumcircle.

21. Right Triangle

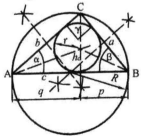

A triangle which contains a right angle (90°) is a right triangle. In the conventional a, b, c labeling of the three sides, the side of length c will represent the hypotenuse.

a) Area:

$$A = \frac{ab}{2} = \frac{ch_c}{2} = \frac{c}{2}\sqrt{pq} = \frac{c}{2}pq^{\frac{1}{2}}$$

where

$$a = \sqrt{pc} = pc^{\frac{1}{2}}$$

$$b = \sqrt{qc} = qc^{\frac{1}{2}}$$

$$h_c = pq$$

b) Perimeter:

$$P = a + b + c$$

where

a, b, c = the sides of the triangle.

c) Radius of inscribed circle:

$$r = \frac{ab}{a+b+c} = s - c$$

d) Radius of circumscribed circle:

$$R = \frac{c}{2}$$

---

22. Ratio of the Sides of a Right Triangle

A ratio is a comparison by division. Each ratio is assigned a name, and these names are called *functions.*

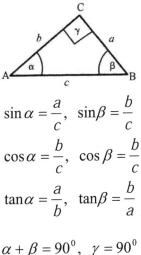

$$\sin\alpha = \frac{a}{c}, \quad \sin\beta = \frac{b}{c}$$

$$\cos\alpha = \frac{b}{c}, \quad \cos\beta = \frac{b}{c}$$

$$\tan\alpha = \frac{a}{b}, \quad \tan\beta = \frac{b}{a}$$

$$\alpha + \beta = 90^0, \quad \gamma = 90^0$$

23. Pythagorean Theorem

The Pythagorean theorem states that in any right triangle, the area of the square of the hypotenuse is equal to the sum of the areas of the squares of the other two sides. It can be used to find an unknown side of a right-angled triangle, or to prove that a given triangle is right angled.

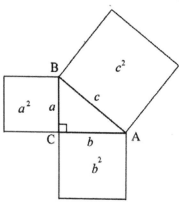

If vertex C is the right angle, we can write this as

$$c^2 = a^2 + b^2$$

24. Equilateral Triangles

A triangle with all three sides of equal length and three
60^0 angles is an equilateral triangle.

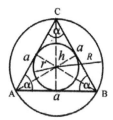

a) Angles:

$$A = B = C = \alpha = 60^0$$

b) Perimeter:

$$P = 3a$$

c) Altitude:

$$h = \frac{a}{2}\sqrt{3}$$

d) Area:

$$A = \frac{a^2}{4}\sqrt{3}$$

e) Radius of inscribed circle:

$$r = \frac{a}{6}\sqrt{3} = \frac{R}{2}$$

f) Radius of circumscribed circle:

$$R = \frac{a}{3}\sqrt{3}$$

25. Isosceles Triangle

A triangle with two sides of equal length is an isosceles triangle.

a) Area:

$$A = \frac{c h_c}{2} = \frac{c}{2}\sqrt{4a^2 - c^2}$$

b) Perimeter:

$$P = 2a + c$$

c) Altitude:

$$h_c = \frac{\sqrt{4a^2 - c^2}}{2} = a\cos\frac{\gamma}{2}$$

d) Radius of inscribed circle:

$$r = \frac{2A}{P}$$

e) Radius of circumscribed circle:

$$R = \frac{a^2 c}{4A}$$

f) Angles:

$$2\alpha + \gamma = 180^0$$

where

α = base angles (congruent)
γ = vertex angle

26. Square

A square is a closed planar quadrilateral with all sides of equal length a, and with four right angles.

a) Perimeter:

$$P = 4a$$

b) Area:

$$A = a^2 = \frac{d^2}{2}$$

c) Radius of inscribed circle:

$$r = \frac{a}{2}$$

d) Radius of circumscribed circle:

$$R = \frac{d}{2}$$

e) Diagonals:

$$d_1 = d_2 = d = a\sqrt{2}$$

27. Rectangle

A rectangle is a closed planar quadrilateral with opposite sides of equal lengths a and b, and with four right angles.

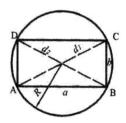

a) Perimeter:

$$P = 2(a + b)$$

b) Area:

$$A = ab$$

c) Diagonals:

$$d_1 = d_2 = d = \sqrt{a^2 + b^2}$$

d) Radius of circumscribed circle:

$$R = \frac{d}{2}$$

28. Parallelogram

A parallelogram is a closed planar quadrilateral whose opposite sides are parallel.

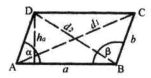

a) Perimeter:

$$P = 2(a + b)$$

b) Area:

$$A = ah_a = ab\sin\alpha$$

c) Diagonals:

$$d_1 = \sqrt{(a + h_a \cot\alpha)^2 + h_a^2} = \sqrt{a^2 + b^2 - 2ab\cos\beta}$$

$$d_2 = \sqrt{(a - h_a \cot a)^2 + h_a^2} = \sqrt{a + b^2 - 2ab\cos\alpha}$$

$$d_1^2 + d_2^2 = 2(a^2 + b^2)$$

29. Rhombus

A rhombus is a closed planar parallelogram with all sides equal.

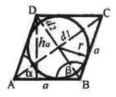

a) Area:

$$A = a^2 \sin\alpha = a^2 \sin\beta = ah_a = \frac{d_1 d_2}{2}$$

b) Diagonals:

$$d_1 = 2a\cos\frac{\alpha}{2}; \ d_2 = 2a\sin\frac{\alpha}{2}$$

$$d_1^2 + d_2^2 = 4a^2$$

c) Radius of inscribed circle:

$$r = \frac{d_1 d_2}{2\sqrt{d_1^2 d_2^2}}.$$

d) Altitude:

$$h_a = a\sin\alpha$$

30. Trapezoid (American definition)

A trapezoid is a quadrilateral with one and only one pair of parallel sides.

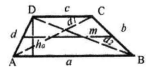

a) Perimeter:

$$P = a + b + c + d$$

b) Area:

$$A = \frac{a+c}{2}h_a = mh_a$$

$$m = \frac{a+c}{2}, \qquad (a \neq c,\ c \parallel a)$$

c) Altitude:

$$h_a^2 = \frac{k \cdot l}{4(a-c)^2}$$

where

$$k = (a+d-c+b) \cdot (d+c+b-a)$$
$$l = (a-d-c+b) \cdot (a+d-c-b)$$

31. Kite

A kite is a closed planar quadrilateral whose two pairs of distinct adjacent sides are equal in length. One diagonal bisects the other. Diagonals intersect at right angles

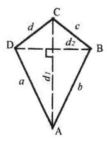

a) Perimeter:

$$P = a + b + c + d$$

where

a, b, c, d = sides of kite, $(a = b,\ c = d)$

b) Area:

$$A = \frac{d_1 d_2}{2}$$

where

d_1, d_2 = diagonals kite, $(d_1 \perp d_2)$

32. Regular Polygons

A regular polygon is a closed plane figure with n sides. If all sides and angles are equivalent, the polygon is called regular

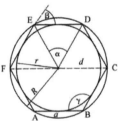

a) Perimeter:

$$P = n \cdot a$$

b) Area:

$$A = \frac{n}{2} R^2 \sin \alpha$$

c) Radius of circumscribed circle:

$$R = \frac{a}{2 \sin \dfrac{\alpha}{2}}$$

d) Radius of inscribed circle:

$$r = \frac{a}{2 \tan \dfrac{\alpha}{2}}$$

e) Central angles:

$$\alpha = \frac{360^0}{n}$$

f) Internal angles:

$$\gamma = 180^0 - \beta = \frac{n-2}{n} \cdot 180^0$$

g) External angles:

$$\beta = \alpha$$

h) Number of diagonals:

$$N = \frac{1}{2} n(n-3)$$

33. Circle

All points on the circumference of a circle are equidistant from its center.

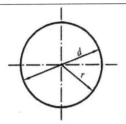

a) Perimeter:

$$P = 2\pi r = \pi d$$

b) Area:

$$A = \frac{\pi}{4} d^2 = \pi r^2$$

34. Sector of a Circle

A sector of a circle of radius r is the interior portion of the circle determined by a central angle α

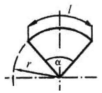

a) Angle:

$$\widehat{\alpha} = \frac{\pi}{180^0} \alpha^0 \ [\text{rad}], \quad \alpha^0 = \frac{\widehat{\alpha}}{2} r^2 \ \big[^0\big]$$

b) Length of arc:

$$l = \frac{\pi}{180^0} r a^0$$

c) Area:

$$A = \frac{\pi}{360^0} r^2$$

35. Segment of a Circle

A segment is a portion of a circle whose upper boundary is a circular arc l and lower boundary is a secant s.

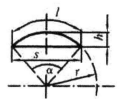

a) Angle:

$$\hat{\alpha} = \frac{\pi}{180^0} \alpha^0 \; [\text{rad}], \quad \alpha^0 = \frac{\hat{\alpha}}{2} r^2 \; \left[^0\right]$$

b) Radius:

$$r = \frac{h}{2} + \frac{s}{8h}$$

c) Secant:

$$s = 2 r \sin \frac{\alpha}{2} = 2\sqrt{h(2r - h)}$$

d) Area:

$$A = \frac{r^2}{2}\left(\hat{\alpha} - \sin\alpha^0\right) \approx \frac{h}{6s}\left(3h^2 + 4s^2\right)$$

e) Height:

$$h = r\left(1 - \cos\frac{\alpha^0}{2}\right) = \frac{s}{2}\tan\frac{\alpha^0}{4}$$

36. Annulus (Circular Ring)

The annulus is the plane area between two concentric circles, making a flat ring.

Area:

$$A = \frac{\pi}{4}\left(D^2 - d^2\right) = \pi(d + w)w$$

37. Ellipse

An ellipse is the locus of a point that moves in such a way that the sum of its distance from two fixed points (the foci) is constant.

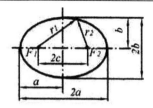

a) Radii:

$$r_1^2 + r_2^2 = 2a$$

b) Area:

$$A = \pi ab$$

c) Perimeter:

$$P \approx 2\pi \sqrt{\frac{1}{2}\left(a^2 + b^2\right)} = \pi(a + b)k$$

a) Radii:

$$r_1^2 + r_2^2 = 2a$$

b) Area:

$$A = \pi ab$$

c) Perimeter:

$$P \approx 2\pi \sqrt{\frac{1}{2}\left(a^2 + b^2\right)} = \pi(a + b)k$$

$$k = 1 + \frac{1}{4}m^2 + \frac{1}{64}m^4 + \frac{1256}{}m^6 + ..., \quad m = \frac{a - b}{a + b}$$

38. Cube

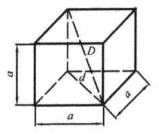

The cube is a regular hexahedron. It is composed of six square planes that meet each other at right angles; it has 12 edges.

a) Area:

$$A = 6a^2$$

b) Volume:

$$V = a^3$$

c) Diagonal of cube:

$$D = a\sqrt{3}$$

d) Diagonal of square:

$$d = a\sqrt{2}$$

39. Cuboid

The cuboid is a solid body composed of three pairs of rectangular planes placed opposite each other and joined at right angles to each other.

Solid Bodies

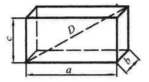

a) Area:

$$A = 2(ab + ac + bc)$$

b) Volume:

$$V = abc$$

c) Diagonal:

$$D = \sqrt{a^2 + b^2 + c^2}$$

40. Right Prism

A right prism is a solid body in which the bases (top and bottom) are right polygons so that the vertical polygons connecting their sides are not only parallelograms, but also right figures.

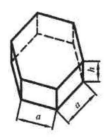

a) Volume:

$$V = A_b h$$

b) Area:

$$A = 2A_b + A_l$$

where

A_b = area of base

A_l = lateral area

41. Pyramid

A pyramid is a solid body whose base is a polygon and whose other planes are all triangles meeting at the apex. A right pyramid has for the base a right polygon.

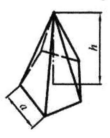

a) Volume:

$$A = A_b + A_l$$

where

A_b = area of base

A_l = the lateral area

---

42. Frustum of Pyramid

Slicing the top off a pyramid creates a frustum of a pyramid. It is determined by the plane of the base and a plane parallel to the base.

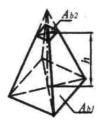

a) Area:

$$A = A_{b1} + A_{b2} + A_l$$

where

A_{b1}, A_{b2} = area of bases

A_l = the lateral area

h = height of pyramid

b) Volume:

$$V = \frac{h}{3}(A_{b1} + A_{b2} + \sqrt{A_{b1} \cdot A_{b2}})$$

43. Cone

A cone is a solid of the form described by the revolution of a right-angled triangle about one of the sides adjacent to the right angle; also called a right cone.

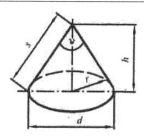

a) Volume:

$$V = A_b h = \frac{\pi}{3} r^2 h$$

b) Area:

$$A = A_b + A_l = \pi \cdot r^2 + \pi \cdot r \cdot s = \pi \cdot r(r + s)$$

where:

A_b = area of base

A_l = the lateral area.

s = slant height

c) Lateral area:

$$A_l = 2\pi \cdot r \cdot h$$

d) Area of base:

$$A_b = \frac{d^2 \pi}{2} = \pi r^2$$

e) Slant height:

$$s = \sqrt{r^2 + h^2}$$

---

f) Vertex angle:

$$\vartheta = 2 \tan^{-1}\left(\frac{r}{h}\right)$$

g) Height:

$$h = \sqrt{s^2 - r^2}$$

44. Frustum of Cone

Slicing the top off a cone creates a frustum of a cone. The plane of the base and a plane parallel to the base determine it.

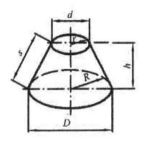

a) Area:

$$A = A_{b1} + A_{b2} + A_l$$
$$A = \pi\left[R^2 + r^2 + (R+r)s\right]$$

b) Area of bases:

$$A_{b1} = \pi R^2, \quad A_{b2} = \pi \cdot r^2$$

c) Lateral area:

$$A_l = \pi \cdot s(R + r)$$

d) Slant height:

$$s = \sqrt{h^2 + (R - r)^2}$$

e) Volume:

$$V = \frac{1}{3}\pi \cdot h \cdot (R^2 + r^2 + R \cdot r)$$

45. Cylinder

A cylinder is a solid body with a circular base and straight sides.

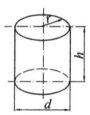

a) Area:

$$A = 2A_b + A_l$$

$$A = 2\pi \cdot r^2 + 2\pi \cdot r \cdot h$$

$$A = 2\pi \cdot r(r + h).$$

b) Area of base:

$$A_b = \frac{\pi \cdot d^2}{4} = \pi \cdot r^2$$

c) Lateral area:

$$A_l = 2\pi \cdot r \cdot h$$

d) Volume:

$$V = \frac{\pi}{4} d^2 h = \pi r^2 h$$

46. Hollow Cylinder

A hollow cylinder is a solid with circular ring bases and straight sides.

a) Volume:

$$V = A_b \cdot h = \frac{\pi}{4} h \cdot \left(D^2 - d^2 \right)$$

where

A_b = annulus area

h = height of cylinder

w = thickness of wall

D, d = outside and inside diameters of the hollow cylinder

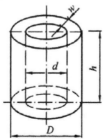

47. Sliced Cylinder

A sliced cylinder is a portion of a circular cylinder cut off by a sloped plane.

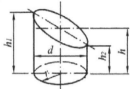

a) Area:

$$A = A_{b1} + A_{b2} + A_l$$

$$A = \pi \cdot r \left[h_1 + h_2 + r + \sqrt{r^2 + \frac{(h_1 + h_2)^2}{4}} \right]$$

b) Lateral area:

$$A_l = \pi \cdot d \cdot h$$

c) Volume:

$$V = \frac{\pi}{4} d^2 h$$

48. Sphere

A sphere is defined as a three-dimensional figure with all of its points equidistant at distance r from its center.

a) Area:

$$A = 4\pi \cdot r^2$$

b) Volume:

$$V = \frac{4\pi \cdot r^3}{3}$$

49. Spherical Cap

A spherical cap is the portion of a sphere cut off by a plane.

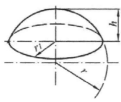

a) Area:

$$A = 2\pi rh = \pi\left(r_1^2 + h^2\right)$$

b) Volume:

$$V = \frac{1}{3}\pi h^2 (3r - h)$$

50. Sector of a Sphere

A sector of a sphere is the part of a sphere generated by a right circular cone that has its vertex at the center of the sphere.

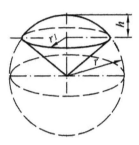

a) Area:

$$A = \pi \cdot r(ah + r_1)$$

b) Volume:

$$V = \frac{2}{3}\pi \cdot r^2 \cdot h$$

where

r = radius of the sphere

r_1 = radius of the base

51. Zone of a Sphere

The zone of a sphere is a portion cut off by two parallel planes.

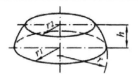

a) Area:

$$A = A_{b1} + A_{b2} + A_l$$
$$A = \pi r_1^2 + \pi r_2^2 + 2\pi \cdot r \cdot h$$
$$A = \pi\left(2rh + r_1^2 + r_2^2\right)$$

where

A_{b1}, A_{b2} = area of bases

A_l = the lateral area zone of a sphere

r_1, r_2 = radii of bases

b) Volume:

$$V = \frac{\pi}{6} h\left(3r_1^2 + 3r_2^2 + h^2\right)$$

52. Torus

A torus is the surface of a three-dimensional figure obtained by rotating a circle about an axis coplanar with the circle and a fixed distance from the origin.

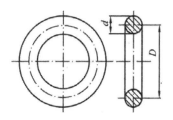

a) Area:

$$A = \pi^2 Dd$$

b) Volume:

$$V = \frac{\pi^2}{4} Dd^2$$

53. Ellipsoid

An ellipsoid is a three-dimensional figure, all planar cross-sections of which are ellipses. Semi-axes: a, b, c $(a \neq b \neq c)$. If two of those are equal, the ellipsoid is a spheroid; if all three are equal, it is a sphere.

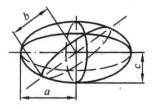

a) Volume:

$$V = \frac{4}{3}\pi \cdot abc$$

where

a, b, c = half axes of ellipsoid

54. Barrel

A barrel is a solid that bulges out in the middle and has circular ends.

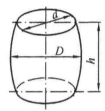

a) Volume:

$$V = \frac{\pi}{12} h \left(2D^2 + d^2 \right)$$

TRIGONOMETRY

Trigonometry is the branch of mathematics concerned with solving triangles, circles, oscillations, and waves using trigonometric ratios, which are seen as properties of triangles rather than of angles. It is absolutely crucial to much of geometry and physics.

This section contains:
1. Fundamentals of Trigonometry
2. Trigonometric Equations
3. Graphs of the Trigonometric Functions

1. Circular and Angular Measures

An angle is formed by two intersecting half lines or by rotating a half line from position OP to its terminal position OR. If the rotation is clockwise, the angle is deemed negative, and if counterclockwise the angle is deemed positive.

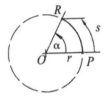

a) Circular measure

The circular measure is the ratio of the arc $PR = s$ to the radius r:

$$\hat{\alpha} = \frac{s}{r} = 1 \text{ (rad)}$$

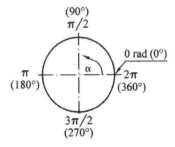

b) Angular measure

The angular degree symbolized by the (°) is a unit of plane angular measure. There are 360 angular degrees in a complete circle. Each degree is divided into 60 minutes and each minute is divided into 60 seconds.

c) Relation between circular and angular measure:

de-grees	0^0	30^0	60^0	90^0	180^0	270^0	360^0
radians	0	$\dfrac{\pi}{6}$	$\dfrac{\pi}{3}$	$\dfrac{\pi}{2}$	π	$\dfrac{3\pi}{2}$	2π
	0	0.52	1.05	1.57	3.14	4.71	6.28

1 rad = 57.2958 degrees

2. Trigonometric Circle

A circle centered in origin O and with radius = 1 is called a trigonometric circle or unit circle.

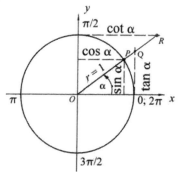

The x-coordinate of point P is called the cosine of α.
The y-coordinate of point P is called the sine of α.
The y-coordinate of point Q is called the tangent of α.
The x-coordinate of point R is called the cotangent of α.

3. **Basic Formulas**

$$\sin^2 \alpha + \cos^2 \alpha = 1$$

$$\tan \alpha = \frac{\sin \alpha}{\cos \alpha}$$

$$\cot \alpha = \frac{\cos \alpha}{\sin \alpha} = \frac{1}{\tan \alpha}$$

$$\cos(-\alpha) = \cos \alpha$$

$$\sin(-\alpha) = -\sin \alpha$$

$$\sin \alpha = \sin(\alpha + 2\pi)$$

$$\cos \alpha = \cos(\alpha + 2\pi)$$

$$1 + \tan^2 \alpha = \sec^2 \alpha$$

$$1 + \cot^2 \alpha = \cos^2 \alpha = \frac{1}{\sin^2 \alpha}$$

If α and α' are supplementary values $(\alpha + \alpha' = \pi)$, then

$$\sin \alpha = \sin(\alpha')$$
$$\cos \alpha = -\cos(\alpha')$$
$$\tan \alpha = -\tan(\alpha')$$
$$\cot \alpha = -\cot(\alpha')$$

If α and α' are complementary values $\left(\alpha + \alpha' = \dfrac{\pi}{2}\right)$,

then

$$\sin\alpha = \cos(\alpha')$$
$$\cos\alpha = \sin(\alpha')$$
$$\tan\alpha = \cot(\alpha')$$
$$\cot\alpha = \tan(\alpha')$$

If α and α' are opposite values $(\alpha + \alpha' = 0)$, then

$$\sin\alpha = -\sin(\alpha')$$
$$\cos\alpha = \cos(\alpha')$$
$$\tan\alpha = -\tan(\alpha')$$
$$\cot\alpha = -\cot(\alpha')$$

4. Trigonometric Ratios for Right Angled Triangles

There are six ratios, defined as follows: three *major* and three *minor.*

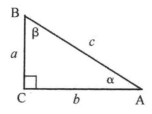

Fundamentals of Trigonometry

Major:	Minor:
$\sin\alpha = \dfrac{\text{opp}}{\text{hyp}} = \dfrac{a}{c}$	$\cot\alpha = \dfrac{\text{adj}}{\text{opp}} = \dfrac{b}{a}$
$\cos\alpha = \dfrac{\text{adj}}{\text{hyp}} = \dfrac{b}{c}$	$\sec\alpha = \dfrac{\text{opp}}{\text{hyp}} = \dfrac{a}{c}$
$\tan\alpha = \dfrac{\text{opp}}{\text{adj}} = \dfrac{a}{b}$	$\csc\alpha = \dfrac{\text{hyp}}{\text{opp}} = \dfrac{c}{a}$

5. Sum and Difference of Functions of Angles

$$\sin\alpha + \sin\beta = 2\sin\frac{\alpha+\beta}{2}\cos\frac{\alpha-\beta}{2}$$

$$\sin\alpha - \sin\beta = 2\cos\frac{\alpha+\beta}{2}\sin\frac{\alpha-\beta}{2}$$

$$\sin\alpha + \sin\beta = 2\sin\frac{\alpha+\beta}{2}\cos\frac{\alpha-\beta}{2}$$

$$\sin\alpha - \sin\beta = 2\cos\frac{\alpha+\beta}{2}\sin\frac{\alpha-\beta}{2}$$

$$\cos\alpha + \cos\beta = 2\cos\frac{\alpha+\beta}{2}\cos\frac{\alpha-\beta}{2}$$

6. Sum and Difference of Angles

$$\sin(\alpha \pm \beta) = \sin\alpha\cos\beta \pm \cos\alpha\sin\beta$$

$$\cos(\alpha \pm \beta) = \cos\alpha\cos\beta \pm \sin\alpha\sin\beta$$

$$\tan(\alpha \pm \beta) = \frac{\tan\alpha \pm \tan\beta}{1 \pm \tan\alpha\tan\beta}$$

$$\cot(\alpha \pm \beta) = \frac{\cot\alpha\cot\beta \mp 1}{\pm\cot\alpha + \cot\beta}$$

7. Double Angle Formulas

$$\sin 2\alpha = 2\sin\alpha\cos\alpha$$

$$\cos 2\alpha = \cos^2\alpha - \sin^2\alpha$$

$$\tan 2\alpha = \frac{2\tan\alpha}{1 - \tan^2\alpha}$$

$$\cot 2\alpha = \frac{\cot^2\alpha - 1}{2\cot\alpha}$$

8. Half Angle Formulas

$$\sin\frac{\alpha}{2} = \sqrt{\frac{1 - \cos\alpha}{2}}$$

$$\cos\frac{\alpha}{2} = \sqrt{\frac{1 + \cos\varepsilon}{2}}$$

Fundamentals of Trigonometry

$$\tan\frac{\alpha}{2} = \sqrt{\frac{1-\cos\alpha}{1+\cos\alpha}}$$

$$\cot\frac{\alpha}{2} = \sqrt{\frac{1+\cos\alpha}{1-\cos\alpha}}$$

9. Functions of Important Angles

α	$(^0)$	0^0	30^0	60^0	90^0	120^0	180^0
	rad	0	$\dfrac{\pi}{6}$	$\dfrac{\pi}{3}$	$\dfrac{\pi}{2}$	$\dfrac{2\pi}{3}$	π
$\sin\alpha$		0	$\dfrac{1}{2}$	$\dfrac{\sqrt{3}}{2}$	1	$\dfrac{\sqrt{3}}{2}$	0
$\cos\alpha$		1	$\dfrac{\sqrt{3}}{2}$	$\dfrac{1}{2}$	0	$-\dfrac{1}{2}$	-1
$\tan\alpha$		0	$\dfrac{\sqrt{3}}{3}$	$\sqrt{3}$	$\pm\infty$	$-\sqrt{3}$	0
$\cot\alpha$		$\pm\infty$	$\sqrt{3}$	$\dfrac{\sqrt{3}}{3}$	0	$-\dfrac{\sqrt{3}}{3}$	$\pm\infty$

10. Solving Trigonometric Equations

Some equations that involve trigonometric functions of an unknown may be readily solved by using simple algebraic ideas, while others may be impossible to solve exactly but only approximately. Here are some methods for solving trigonometric equations:

  a) Find the solution to the equation, and reduce to base equation:

Example:

$$\tan\left(x - \frac{\pi}{2}\right) = \tan 2x \qquad (1)$$

Solution:
Reduce equation (1) to the base equation

$$\left(x - \frac{\pi}{2}\right) = 2x + k\pi$$

$$-x = \frac{\pi}{2} + k\pi$$

It follows that

$$x = -\frac{\pi}{2} + k\pi$$

  b) Find the solution to the equation using factorization:

Example:

$$2\cos^2 x - 5\cos x + 2 = 0 \qquad (1)$$

Solution:

After factorization, Equation (1) has the form

$$(2\cos x - 1)(\cos x - 2) = 0$$

The roots of the equation are:

$$2\cos x - 1 = 0$$
$$2\cos x = 1$$
$$\cos x = \frac{1}{2}, \text{ and}$$
$$\cos x - 2 = 0$$
$$\cos x = 2$$

Remember that the range for $\cos x$ is

$$\{y | -1 \le y \le 1, \text{ y is real}\}$$

That is, y is between (-1) and 1, inclusive $\cos x \ne 2$.
Hence, the root that satisfies Equation (1) is

$$x = \frac{\pi}{3} + k\pi, \ k \in \text{integer}$$

   c) Find the solution to the equation using an additional
      unknown:

Example:

$$2\sin^2(2x) + \sin(2x) - 1 = 0$$

Solution:
Let $u = \sin(2x)$

$$2u^2 + u - 1 = 0$$

$$u_{1,2} = -\frac{b \pm \sqrt{b^2 - 4ac}}{2a}$$

$$u_1 = \frac{1}{2}, \ u_2 = -1$$

Substitute:

$$\sin(2x) = \frac{1}{2}, \text{ and } \sin(2x) = -1.$$

$$\sin(2x) = \sin\frac{\pi}{6} \text{ or } \sin(2x) = \sin\left(-\frac{\pi}{6}\right)$$

$$2x = \frac{\pi}{6} + 2k\pi \text{ or } 2x = \pi - \frac{\pi}{6} + 2k\pi$$

$$X = \begin{cases} \dfrac{\pi}{12} + k\pi, \ \dfrac{5\pi}{12} + k\pi, \ -\dfrac{\pi}{4} + k\pi, \text{ or} \\ \dfrac{3\pi}{2} + k\pi, \ k \in \text{integer} \end{cases}$$

11. Verifying Trigonometric Identities

Example:
Verify identities

Trigonometric Equations

$$\frac{1 + \tan x}{1 + \cot x} = \frac{\sin x}{\cos x}$$

Solution:

Identity used:

$$\tan x = \frac{\sin x}{\cos x}; \ \cot x = \frac{1}{\tan x}$$

Simplify principal numerator and principal denominator of the left term.

$$\frac{1 + \dfrac{\sin x}{\cos x}}{1 + \dfrac{\cos x}{\sin x}} = \frac{\sin x}{\cos x}$$

Divide the principal denominator into the principal numerator of the left term.

$$\frac{\cos x + \sin x}{\cos x} \cdot \frac{\sin x}{\sin x + \cos x} = \frac{\sin x}{\cos x}$$

Reduce the left term by the factor $\sin x + \cos x$

$$\frac{\sin x}{\cos x} = \frac{\sin x}{\cos x}$$

Hence, the identities are correct.

12. Graphs of the Sine and Cosine Functions

$$y = \sin x$$
$$y = \cos x \qquad \text{for} \quad (-\frac{\pi}{2} \le x \le 2\pi)$$

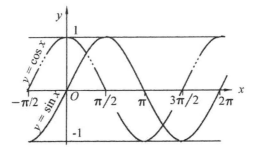

Domain: all real numbers.

Range: $-1 < y < 1$

Period: 2π

13. Graphs of the Tangent and Cotangent Functions

$$y = \tan x$$
$$y = \cot x \qquad \text{for} \quad (-\pi < x < \pi)$$

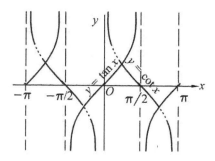

Domain: x is all real numbers except the tangent function $\dfrac{\pi}{2} + k\pi$, and cotangent function $k\pi$, (asymptotes occur here).

Range: all real numbers

Period: π

ANALYTICAL GEOMETRY

Analytic geometry, also called coordinate geometry, is the study of geometry using the principles of algebra. The Cartesian coordinate system is usually applied to manipulate equations for planes, lines, curves, and circles, often in two but sometimes in three dimensions of measurement.

This section contains the most frequently used formulas, rules, and definitions relating to the following:

1. Points and lines
2. Circles
3. Ellipses
4. Parabolas
5. Hyperbolas
6. Polar Coordinates
7. Solid Analytical Geometry
8. Planes
9. The Straight Line in Space
10. Surfaces

---

1. Distance between Two Points

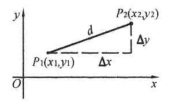

The distance between two points $P_1(x_1, y_1)$
and $P_2(x_2, y_2)$ is defined by the formula

$$d = \sqrt{(x_2 - x_1)^2 + (y_2 - y_1)^2}$$

where

$$\Delta x = x_2 - x_1$$
$$\Delta y = y_2 - y_1$$

2. Point of Division

The point of division is the point $P(x, y)$ which divides a
line segment $P1(x1, y1)$, $P2(x2, y2)$ in a given ratio,

$$\lambda = \frac{P_1 P}{P P_2}$$

Point P has the coordinates

$$x = \frac{x_1 + \lambda x_2}{1 + \lambda}, \quad y = \frac{y_1 + \lambda y_2}{1 + \lambda}$$

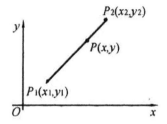

If $P(x, y)$ is the midpoint of line $P_1(x_1, y_1), P_2(x_2, y_2)$, $\lambda = 1$, then point P has the coordinates

$$x = \frac{x_1 + x_2}{2}, \quad y = \frac{y_1 + y_2}{2}$$

3.   Inclination and Slope of a Line

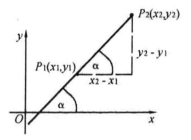

a) Inclination

The inclination of a line not parallel to the x-axis is defined as the smallest positive angle measured from the positive direction of the x-axis in a counterclockwise direction to the line. If the line is parallel to the x-axis, its inclination is defined as zero.

b) Slope

The slope of a line passing through two points $P_1(x_1, y_1)$ and $P_2(x_2, y_2)$ is

$$m = \tan \alpha = \frac{y_2 - y_1}{x_2 - x_1}$$

4. Parallel and Perpendicular Lines

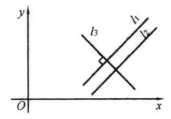

If line l_1 is parallel with line l_2, then their slopes are equal:

$$m_1 = m_2$$

If line l_1 and l_3 are perpendicular, the slope of one of the lines is the negative reciprocal of the slope of the other line.

If m_1 is the slope of l_1 and m_3 is the slope of l_3, then

$$m_1 = -\frac{1}{m_3}, \quad \text{or} \quad m_1 \cdot m_3 = -1$$

5.   Angle Between Two Intersection Lines

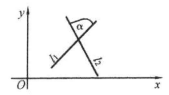

Angle α, measured in a positive direction counterclockwise from line l_1, whose slope is m_1 to the line l_2, whose slope is m_2, is

$$\tan \alpha = \frac{m_2 - m_1}{1 + m_1 m_2}$$

6.   Triangle

The area of a triangle in terms of the vertices is

$$A = \frac{1}{2}\left(x_1 y_2 + x_2 y_3 + x_3 y_1 - x_3 y_2 - x_2 y_1 - x_1 y_3\right)$$

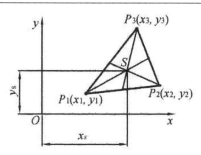

The coordinates of the centroid S (center of gravity) of the triangle are

$$x_s = \frac{x_1 + x_2 + x_3}{3}$$

$$y_s = \frac{y_1 + y_2 + y_3}{3}$$

7. The Equation for a Straight Line through a Point

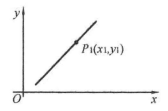

A straight line is completely determined if its gradient is known and a point $P(x_1, y_1)$ is given through which the line must pass:

$$y - y_1 = m(x - x_1)$$

8. Slope-Intercept Form

A straight line is defined if its slope (gradient) m is known and the y-intercept is $(0, b)$. Its equation is

$$y = mx + b$$

9. Equation for a Straight Line through Two Points

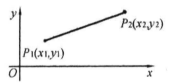

The equation of a straight strength line through two defined points $P_1(x_1, y_1)$, and $P_2(x_2, y_2)$ is

$$\frac{y - y_1}{x - x_1} = \frac{y_1 - y_2}{x_1 - x_2}$$

10. Intercept Form Equation of the Straight Line

$$\frac{x}{a} + \frac{y}{b} = 1$$

---

where
> a = x-intercept
> b = y-intercept

11. General Form of an Equation of a Straight Line

$$Ax + By + C = 0$$

where
> A, B and C are arbitrary constants.

For an equation in this form, the slope m and y-intercept b are

$$m = -\frac{A}{B}$$

$$b = -\frac{C}{B}$$

12. Normal Equation of a Straight Line

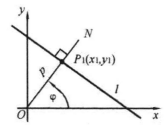

A straight line is defined if the length of the perpendicular (p) from origin (0.0) to the line is known, and if the angle (φ) which this perpendicular makes with the x-axis, is known.

The normal form of the equation of the straight line is

$$x \cos\varphi + y \sin\varphi - p = 0$$

The normal form of equation $Ax + By + C = 0$ is

$$\frac{A}{\pm\sqrt{A^2 + B^2}} x + \frac{B}{\pm\sqrt{A^2 + B^2}} y + \frac{C}{\pm\sqrt{A^2 + B^2}} = 0$$

where

$$\cos\varphi = \frac{A}{\pm\sqrt{A^2 + B^2}}$$

$$\sin\varphi = \frac{B}{\pm\sqrt{A^2 + B^2}}$$

$$-p = \frac{C}{\pm\sqrt{A^2 + B^2}}$$

13. Distance From a Line to a Point

The distance from a line l to a point P_1 (x_1, y_1) is perpendicular distance d.

Since the coordinates of point P_1 (x_1, y_1) satisfy the equation for l_1,

$$x_1 \cos\varphi + y_1 \sin\varphi - (p + d) = 0,$$

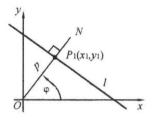

solving for d,

$$d = x_1 \cos \varphi + y_1 \sin \varphi - p \,,$$

or

$$d = \frac{\left| Ax_1 + By_1 + C \right|}{\sqrt{A^2 + B^2}} \,.$$

14. Circles

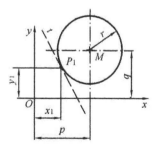

A circle is represented by an equation of the second degree. A circle is completely defined if its center $M\,(p, q)$ and radius r are known.

---

a) The equation of a circle:

$$(x-p)^2 + (y-q)^2 = r^2$$

If the center of a circle is at the origin the equation becomes

$$x^2 + y^2 = r^2$$

The general equation of a circle is

$$x^2 + y^2 + Dx + Ey + F = 0, \text{ or}$$

$$\left(x + \frac{D}{2}\right)^2 + \left(y + \frac{E}{2}\right)^2 = \frac{D^2 + E^2 - 4F}{4}$$

The center of the circle is at the point $M\left(-\dfrac{D}{2}, -\dfrac{F}{2}\right)$,

and the radius of circle is

$$r = \frac{1}{2}\sqrt{D^2 + E^2 - 4F}$$

If $D^2 + E^2 - 4F > 0,$ the circle is real.

If $D^2 + E^2 - 4F < 0,$ the circle is imaginary.

If $D^2 + E^2 - 4F = 0,$ there is no circle.

b) The tangent t at point $P_1(x_1, y_1)$:

$$y = \frac{r^2 - (x - p)(x_1 - p)}{y_1 - q} + q$$

15. Ellipses

An ellipse is a curve in which the sum of the distances from any point on the curve to two fixed points is constant. The two fixed points are called foci (plural of focus).

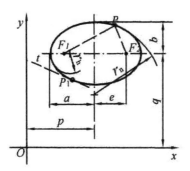

a) The equation of an ellipse:

$$\frac{(x - p)^2}{a^2} + \frac{(y - q)^2}{b^2} - 1 = 0$$

If the center is at the origin, the equation becomes

$$\frac{x^2}{a^2} + \frac{y^2}{b^2} = 1$$

In either case, the general form of the equation of the ellipse is

$$Ax^2 + By^2 + Dx + Ey + f = 0$$

b) Eccentricity:
$$e = \sqrt{a^2 - b^2}, \quad (a > b)$$

c) Vertex radii:
$$r_h = \frac{b^2}{a}, \quad r_n = \frac{a^2}{b}$$

d) Basic property:

$$\overline{F_1 P} + \overline{F_2 P} = 2a$$

where
$$F_1, F_2 = \text{focal points}$$

e) The equation of a tangent t at point $P_1(x_1, y_1)$:

$$y = -\frac{b^2}{a^2} \cdot \frac{(x_1 - p)(x - x_1)}{y_1 - q} + y_1$$

16. Parabolas

A parabola is the set of all points in a plane equidistant from a given line L (the conic section directrix) and a given point F not on the line (the focus). The focal parameter (i.e., the distance between the directrix and focus) is therefore given as p.

The surface of revolution obtained by rotating a parabola about its axis of symmetry is called a <u>paraboloid</u>.

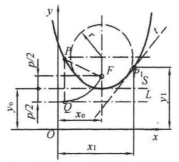

L = directrix S = tangent at the vertex

a) The equation of a parabola:

$$(x - x_0)^2 = 2p(y - y_0) \text{ or}$$
$$(x - 2)^2 = -2p$$

b) Basic equation:

$$y = ax^2 + bx + c = 0$$

c) Vertex radius:

$$r = p$$

d) Basic property:

$$\overline{PF} = \overline{PQ}, \; \frac{\overline{PF}}{\overline{PQ}} = 1 = e$$

e) Equation of a tangent at point $P1$ $(x1, y1)$:

$$y = \frac{2(y_1 - y_0)(x - x_1)}{x_1 - x_0} + y_1$$

17. Hyperbolas

A hyperbola is the set of all points $P(x, y)$ in the plane, the difference of whose distances from two fixed points $F1$ and $F2$ is some constant. The two fixed points are called the foci.

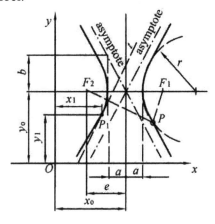

Hyperbolas

---

a) Equation of a hyperbola:

$$\frac{(x - x_0)^2}{a^2} - \frac{(y - y_0)^2}{b^2} - 1 = 0$$

If the point of intersection of asymptotes is at the origin, the equation is

$$\frac{x^2}{a^2} - \frac{y^2}{b^2} - 1 = 0$$

b) Basic equation:

$$Ax^2 + By^2 Cx + Dy + E = 0$$

c) Eccentricity:

$$e = \sqrt{a^2 + b^2}$$

d) The equation of asymptotes:

$$y = \pm\frac{b}{a}x.$$

e) The equation of a tangent at point $P1$ $(x1, y1)$:

$$y = \frac{b^2}{a^2}\frac{(x_1 - x_0)(x - x_1)}{y_1 - y_0} + y_1$$

f) Vertex radius:

$$r = \frac{b^2}{a}$$

18. Polar Coordinates

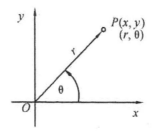

Let x and y be Cartesian axes in the plane and let P be a point in the plane other than the origin. The polar coordinates of point P are r (the radial coordinate) and θ (the angular coordinate, often called the polar angle), and they are defined in terms of Cartesian coordinates by

$$x = r\cos\theta$$

$$y = r\sin\theta$$

where
$\qquad r =$ the radial distance ($r = OP > 0$)
$\qquad \theta =$ the counterclockwise angle from the x-axis

In terms of x and y they are

$$r = \sqrt{x^2 + y^2}$$

$$\theta = \tan^{-1}\left(\frac{y}{x}\right)$$

19. Cartesian Coordinates

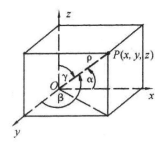

$$OP = \rho, \qquad \rho^2 = x^2 + y^2 + z^2$$

$$x = \rho \cos\alpha, \; y = \rho \cos\beta, \; z = \rho \cos\gamma$$

$$\cos^2\alpha + \cos^2\beta + \cos^2\gamma = 1$$

$$\cos\alpha = \frac{x}{\rho}, \; \cos\beta = \frac{y}{\rho}, \; \cos\gamma = \frac{z}{\rho}, \; \text{or}$$

$$\cos\alpha = \frac{x}{\sqrt{x^2 + y^2 + z^2}}$$

$$\cos\beta = \frac{y}{\sqrt{x^2 + y^2 + z^2}}$$

$$\cos\gamma = \frac{z}{\sqrt{x^2 + y^2 + z^2}}$$

20. Distance between Two Points

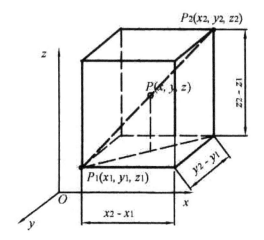

a) Distance between two points $P1$ and $P2$:

$$d = \sqrt{\left(x_2 - x_1\right)^2 + \left(y_2 - y_1\right)^2 + \left(z_2 - z_1\right)^2}$$

b) Point of division

If the point P (x, y, z) divides the line $P1(x1, y1, z1)$ to

$P2$ $(x2, y2, z2)$ in the ratio $\dfrac{P_1P}{PP_2} = \dfrac{r}{1}$, then

$$x = \frac{x_1 + rx_2}{1+r}, \quad y = \frac{y_1 + ry_2}{1+r}, \quad z = \frac{z_1 + rz_2}{1+r}$$

c) Direction of a line
The direction cosines of P_1P_2 are

$$\cos \alpha = \frac{x_2 - x_1}{\sqrt{(x_2 - x_1)^2 + (y_2 - y_1)^2 + (z_2 - z_1)}}$$

$$\cos \beta = \frac{y_2 - y_1}{\sqrt{(x_2 - x_1)^2 + (y_2 - y_1)^2 + (z_2 - z_1)}}$$

$$\cos \lambda = \frac{z_2 - z_1}{\sqrt{(x_2 - x_1)^2 + (y_2 - y_1)^2 + (z_2 - z_1)}}$$

21. Angle between Two Lines

The angle between two lines that do not meet is defined as the angle between two intersecting lines, each of which is parallel to one of the given lines.

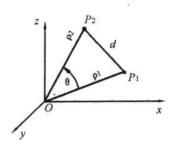

If OP_1 and OP_2 are two lines through the origin parallel to the two given lines, and θ is the angle between the lines, from triangle OP_1P_2, by the law of cosines law,

$$\cos\theta = \frac{\rho_1^2 + \rho_2^2 - d^2}{2\rho_1\rho_2}$$

now

$$\rho_1^2 = x_1^2 + y_1^2 + z_1^2$$
$$\rho_2^2 = x_2^2 + y_2^2 + z_2^2$$
$$d^2 = (x_2 - x_1)^2 + (y_2 - y_1)^2 + (z_2 - z_1)^2$$

Substituting and simplifying

$$\cos\theta = \frac{x_1 x_2 + y_1 y_2 + z_1 z_2}{\rho_1\rho_2} \text{ but,}$$

$$\cos\alpha_1 = \frac{x_1}{\rho_1}, \ \cos\alpha_2 = \frac{x_2}{\rho_2}$$

$$\cos\beta_1 = \frac{y_1}{\rho_1}, \ \cos\beta_2 = \frac{y_2}{\rho_2}$$

$$\cos\gamma_1 = \frac{z_1}{\rho 1}, \ \cos\gamma_2 = \frac{z_2}{\rho_2}$$

Hence,

$$\cos\theta = \cos\alpha_1 \cos\alpha_2 + \cos\beta_1 \cos\beta_2 + \cos\gamma_1 \cos\gamma_2$$

---

22. Every Plane

Every plane is represented by an equation of the first degree in one or more of the variables x, y, z.

a) The equation of a plane:

$$Ax + By + Cz + D = 0, \qquad [(A,B,C) \neq 0]$$

b) The equation of a system of planes passing through a point (x_0, y_0, z_0):

$$A(x - x_0) + B(y - y_0) + C(z - z_0) = 0$$

23. Line Perpendicular to Plane

A line will be perpendicular to a plane $Ax + By + Cz + D = 0$ if and only if the direction numbers a, b, c of the line are proportional to the coefficients of x, y, z in the equation of the plane. Hence:

$$\frac{a}{A} = \frac{b}{B} = \frac{c}{C}, \qquad (a,b,c,A,B,C) \neq 0$$

24. Parallel and Perpendicular Planes

a) Given two planes

$$A_1 x + B_1 y + C_1 z + D_1 = 0, \text{ and}$$
$$A_2 x + B_2 y + C_2 z + D_2 = 0,$$

The planes are parallel if and only if the coefficients of x, y, z, are proportional. Hence,

$$\frac{A_1}{A_2} = \frac{B_1}{B_2} = \frac{C_1}{C_2}$$

b) Two planes are perpendicular if

$$A_1 A_2 + B_1 B + C_1 C_2 = 0$$

25. Distance of a Point from a Plane

The distance between a point $P_1(x_1, y_1, z_1)$ and a plane $Ax + By + Cz + D = 0$ is

$$d = \left| \frac{Ax_1 + By_1 + Cz_1 + D}{\sqrt{A^2 + B^2 + C^2}} \right|$$

26. Normal Form

The normal form of the equation of a plane is

$$x \cos \alpha + y \cos \beta + z \cos \gamma - p = 0$$

where

$p =$ the perpendicular distance from the origin to the plane

$a, \beta, \gamma =$ the direction angles of that perpendicular distance

The normal form of the equation of the plane
$Ax + By + Cz + D = 0$ is

$$\frac{Ax + By + Cz + D}{\pm\sqrt{A^2 + B^2 + C^2}} = 0$$

The sign of the radical is taken opposite to that of D so that the normal distance p will be positive.

27. Intercept Form
The intercept form equation of a plane is

$$\frac{x}{a} + \frac{y}{b} + \frac{z}{c} = 1$$

where

a, b, c = the x, y, z intercepts respectively.

28. Angle between Two Planes

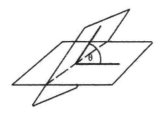

The angle between two planes

$$A_1 x + B_1 y + C_1 z + D_1 = 0, \text{ and}$$
$$A_2 x + B_2 y + C_2 z + D_2 = 0$$

is determined by

$$\cos\theta = \frac{A_1 A_2 + B_1 B_2 + C_1 C_2}{\sqrt{A_1^2 + B_1^2 + C_1^2}\sqrt{A_2^2 + B_2^2 + C_2^2}}$$

29. The Straight Line in Space

The line of intersection of the two planes

$$A_1 x + B_1 y + C_1 z + D_1 = 0, \text{ and}$$
$$A_2 x + B_2 y + C_2 z + D_2 = 0$$

is a straight line in space.

30. Parametric Form Equations of a Line

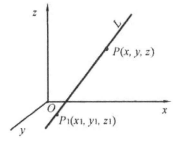

$$x = x_1 + \lambda \cos \alpha$$
$$y = y_1 + \lambda \cos \beta$$
$$z = z_1 + \lambda \cos \gamma$$

or

$$x = x_1 + a\lambda, \quad y = y_1 + b\lambda, \quad z = z_1 + c\lambda$$

where

α, β, γ = the direction angles of the line L

a, b, c = the direction numbers of the line L

λ = the variable length P_1P

31. Symmetric Form Equations of a Line

The equations of the line passing through point $P_1(x_1, y_1, z_1)$ have the form

$$\frac{x - x_1}{\cos \alpha} = \frac{y - y_1}{\cos \beta} = \frac{z - z_1}{\cos \beta \gamma},$$

or

$$\frac{x - x_1}{a} = \frac{y - y_1}{b} = \frac{z - z_1}{c}$$

where

α, β, γ = the direction angles of the line,

a, b, c = the direction numbers of the line

---

32. Two Points Form Equations of a Line

The equations of the straight line through points
$P_1(x_1, y_1, z_1)$ and $P_2(x_2, y_2, z_2)$ are

$$\frac{x = x_1}{x_2 - x_1} = \frac{y - y_1}{y_2 - y_1} = \frac{z - z_1}{z_2 - z_1}.$$

33. Relative Directions of a Line and Plane

A line whose direction numbers are a, b, c and the plane
$Ax + By + Cz + D = 0$ are

a) Parallel when and only when

$$Aa + Bb + Cc = 0 \text{, and}$$

b) Perpendicular when and only when

$$\frac{A}{a} = \frac{B}{b} = \frac{C}{c}.$$

34. The Sphere

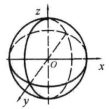

The sphere is a three-dimensional surface, all points of which are equidistant from a fixed point called the center. The equation of a sphere with center at (0, 0, 0) and radius r is

$$x^2 + y^2 + z^2 = r^2$$

If the center of the sphere is at (h, k, j) the equation has the form

$$(x - h)^2 + (y - k)^2 + (z - j)^2 = r^2$$

35. The Ellipsoid

The ellipsoid is a three-dimensional surface, all plane sections of which are ellipses or circles.

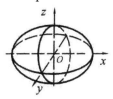

The equation of a ellipsoid with center at (0, 0, 0) and a, b, and c are unequal is

$$\frac{x^2}{a^2} + \frac{y^2}{b^2} + \frac{z^2}{c^2} = 1$$

If $a \neq b$, but $b = c$, the ellipsoid is an ellipsoid of revolution.

If the center of the ellipsoid is (outside of origin) at (h, k, j) and its axes are parallel to the coordinate axes, the equation has the form,

$$\frac{(x-h)^2}{a^2} + \frac{(y-k)^2}{b^2} + \frac{(z-j)^2}{c^2} = 1$$

If the center of the ellipsoid is at the origin, this equation becomes

$$\frac{x^2}{a^2} + \frac{y^2}{b^2} + \frac{z^2}{c^2} = 1$$

36. Hyperboloid

A hyperboloid is a quadric surface generated by rotating a hyperbola around its main axis.

a) Hyperboloid of one sheet:

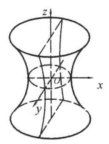

If the equation has the sign of one variable changed, as in

$$\frac{x^2}{a^2} + \frac{y^2}{b^2} - \frac{z^2}{c^2} = 1,$$

the surface is called a hyperboloid of the sheet.
If $a = b$, the surface is a hyperboloid of revolution of one sheet.

b) Hyperboloid of two sheets:

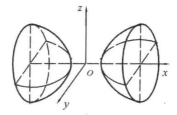

The equation of a hyperboloid of two sheets is

$$\frac{x^2}{a^2} - \frac{y^2}{b^2} - \frac{z^2}{c^2} = 1$$

37. Elliptic Paraboloid

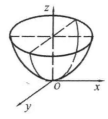

This is the locus of an equation of the form

$$\frac{x^2}{a^2} + \frac{y^2}{b^2} = 2cz$$

The section by a plane $z = k$ is an ellipse that increases in size as the cutting plane recedes from the xy-plane. If $c > 0$, the surface lies wholly above the xy-plane. If $c < 0$, the surface lies wholly below the xy-plane. If $a = b$, the surface is a surface of revolution.

38. Hyperbolic Paraboloid

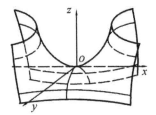

This is the locus of an equation of the form

$$\frac{x^2}{a^2} - \frac{y^2}{b^2} = 2cz, \qquad (c > 0)$$

---

39. Cylindrical Surface

A cylindrical surface is generated by a straight line that moves along a fixed curve and remains parallel to a fixed straight line. The fixed curve is called the *directrix* of the surface and the moving line is the *generatrix* of the surface.

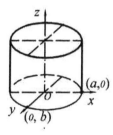

If the directrix is the ellipse for which the standard form for the equation is $b^2x^2 + a^2y^2 = a^2b^2$, the equation of the cylinder is

$$\frac{x^2}{a^2} + \frac{y^2}{b^2} = 1$$

MATHEMATICS OF FINANCE

Financial mathematics is the application of mathematical methods to the solution of problems in finance.

Many people are in the dark when it comes to applying math to practical problem solving. This section will show you how to do the math required to figure out a home mortgage, automobile loan, the present value of an annuity, to compare investment alternatives, and much more.

This section contains formulas, definitions and some examples regarding:

1. Simple interest
2. Compound interest
3. Annuity
4. Amortization

---

1. Simple Interest

Interest is the fee paid for the use of someone else's money. Simple interest is interest paid only on the amount deposited and not on past interest.
The formula for simple interest is

$$I = P \cdot r \cdot t$$

where

I = interest
P = principal
r = interest rate in percent / year
t = time in years

Example:
Find the simple interest for $1500 at 8% for 2 years.

Solution:

P = $1,500, r = 8% = 0.08, and t = 2 years
$I = P \cdot r \cdot t = (1500)(0.08)(2) = 240$ or $240

a) Future value

If P dollars are deposited at interest rate r for t years, the money earns interest. When this interest is added to the initial deposit deposit, the total amount in the account is

$$A = P + I = P + Ptr = P(1 + rt)$$

This amount is called the future value or maturity value.

Example:
Find the maturity value of $10,000 at 8% for 6 months.

Solution:
$P = \$10,000$, $r = 8\% = 0.08$, $t = 6/12 = 0.5$ years

The maturity value is
$A = P(1 + rt) = 10,000[1 + 0.08(0.5)] = 10400$, or
$10,400

2. Compound Interest

Simple interest is normally used for loans or investment of a year or less. For longer periods, compound interest is used.

The compound amount at the end of t years is given by the compound interest formula,

$$A = P(1 + i)^n$$

where

> i = interest rate per compounding period ($i = \dfrac{r}{m}$)
>
> n = number of conversion periods for t years
> ($n = mt$)
>
> A = compound amount at the end of n conversion period
>
> P = principal
>
> r = nominal interest per year
>
> m = number of conversion periods per year
>
> t = term (number of years)

Example:
Suppose $15,000 is deposited at 8% and compounded annually for 5 years. Find the compound amount.

Solution:
$$P = \$15{,}000, \ r = 8\% = 0.08, \ m = 1, \ n = 5$$

$$A = P(1 + i)^n = 1500\left[1 + \left(\frac{0.08}{1}\right)\right]^5 = 15000 \cdot [1.08]^5$$

$$= 22039.92, \text{ or } \$22{,}039.92$$

a) Continuous compound interest
The compound amount A for a deposit of P at interest rate r per year compounded continuously for t years is given by

$$A = Pe^{rt}$$

where
  P = principal
  r = annual interest rate compounded continuously
  t = time in years
  A = compound amount at the end of t years.
  e = 2.7182818

b) Effective rate
The effective rate is the simple interest rate that would produce the same accumulated amount in one year as the nominal rate compounded m times a year.

The formula for effective rate of interest is

$$r_{eff} = \left(1 + \frac{r}{m}\right)^m - 1$$

where

r_{eff} = effective rate of interest

r = nominal interest rate per year

m = number of conversion periods per year

Example:
Find the effective rate of interest corresponding to a nominal rate of 8% compounded quarterly.

Solution:
$r = 8\% = 0.08$, $m = 4$, then

$$r_{eff} = \left(1 + \frac{r}{m}\right)^m - 1 = \left(1 + \frac{0.08}{4}\right)^4 - 1 = 0.082432,$$

so the corresponding effective rate on this case is 8.243% per year.

c) Present value with compound interest
The principal P, is often referred to as the present value, and the accumulated value A, is called the future value since it is realized at a future date. The present value is given by

---

$$P = \frac{A}{(1+i)^n} = A(1+i)^{-n}$$

Example:

How much money should be deposited in a bank paying interest at the rate of 3% per year compounding monthly so that at the end of 5 years the accumulated amount will be $15,000?

Solution:

Here:

- nominal interest per year $r = 3\% = 0.03$,
- number of conversion per year $m = 12$,
- interest rate per compounding period $i = 0.03/12 = 0.0025$,
- number of conversion periods for t years $n = (5)(12) = 60$,
- accumulated amount $A = 15,000$

$$P = A(1+i)^{-n} = 15,000(1 + 0.0025)^{-60}$$
$$P = 12,913.03, \text{ or } \$12,913$$

3. Annuities

An annuity is a sequence of payments made at regular time intervals. This is the typical situation in finding the relationship between the amount of money loaned and the size of the payments.

---

a) Present value of annuity

The present value P of an annuity of n payments of R dollars each, paid at the end of each investment period into an account that earns interest at the rate of i per period, is

$$P = R\left[\frac{1-(1+i)^{-n}}{i}\right]$$

where

P = present value of annuity

R = regular payment per month

n = number of conversion periods for t years

i = annual interest rate

Example:

What size loan could Bob get if he can afford to pay $1,000 per month for 30 years at 5% annual interest?

Solution:

Here: $R = 1,000$, $i = 0.05/12 = 0.00416$, $n = (12)(30) = 360$.

$$P = R\left[\frac{1-(1+i)^{-n}}{i}\right] = 1000\left[\frac{1-(1+0.00416)^{-360}}{0.00416}\right]$$

$P = 186579.61$, or

$186,576.61

Under these terms, Bob would end up paying a total of
$360,000, so the total interest paid would be
$360,000 - $186,579,61 = $173,420.39.

b) Future value of an annuity
The future value S of an annuity of n payments of R
dollars each, paid at the end of each investment period
into an account that earns interest at the rate of i per
period, is

$$S = R\left[\frac{(1+i)^n - 1}{i}\right]$$

Example:
Let us consider the future value of $1,000 paid at the end
of each month into an account paying 8% annual interest
for 30 years. How much will accumulate?

Solution:
This is a future value calculation with $R=1,000$,
$n = 360$, and $i = 0.05/12 = 0.00416$. This account
will accumulate as follows:

$$S = R\left[\frac{(1+i)^n - 1}{i}\right] = 1000\left[\frac{(1+0.00416)^{360} - 1}{0.00416}\right]$$

$S = 831028.59$, or $831,028.59$

Note: This is much larger than the sum of the payments,
since many of those payments are earning interest for
many years.

---

4. Amortization of Loans

The periodic payment Ra on a loan of P dollars to be amortized over n periods with interest charge at the rate of i per period is

$$R_a = \frac{Pi}{1-(1+i)^{-n}}$$

Example:

Bob borrowed \$120,000 from a bank to buy the house. The bank charges interest at a rate of 5% per year. Bob has agreed to repay the loan in equal monthly installments over 30 years. How much should each payment be if the loan is to be amortized at the end of the time?

Solution:

This is a periodic payment calculation with $P = 120,000$, $i = 0.05/12 = 0.00416$, and $n = (30)(12) = 360$

$$R_a = \frac{Pi}{1-(1+i)^{-n}} = \frac{(120000)(0.00416)}{1-(1.00416)^{-360}} = 643.88$$

or \$643.88.

5. Sinking Fund Payment

The Sinking Fund calculation is used to calculate the periodic payments that will accumulate by a specific future date to a specified future value, so that investors can be certain that the funds will be available at maturity.

The periodic payment R required to accumulate a sum of S dollars over n periods, with interest charged at the rate or i per period, is

$$R = \frac{iS}{(1+i)^n - 1}$$

where

S = the future value

i = annual interest rate

n = number of conversion periods for t years

CALCULUS

Calculus is a branch of mathematics developed from algebra and geometry and built on two major complementary ideas.

One concept is *differential calculus*. It studies rates of change, such as how fast an airplane is going at any instant after take-off, the acceleration and speed of a free-falling body at a particular moment, etc.

The other key concept is *integral calculus*. It studies the accumulation of quantities, such as areas under a curve, linear distance traveled, or volume displaced.

Integral calculus is the mirror image of differential calculus.

This section contains:

1. Limits
2. Derivatives
3. Application of Derivatives
4. Integration
5. Basic Integrals
6. Application of Integration

---

1. Limits

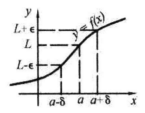

If the value of the function $y = f(x)$ gets arbitrarily close to L as x approaches the point a, then we say that the limit of the function as x approaches a is equal to L. This is written as

$$\lim_{x \to a} f(x) = L$$

2. Rule for Limits

Let u and v be functions such that

$$\lim_{x \to a} u(x) = A \quad \text{and} \quad \lim_{x \to a} v(x) = B$$

1) $$\lim_{x \to a}[ku(x) \pm hv(x)] = k\lim_{x \to a} u(x) \pm h\lim_{x \to a} v(x) = A \pm B$$

2) $$\lim_{x \to a}[u(x) \cdot v(x)] = \left[\lim_{x \to a} u(x)\right] \cdot \left[\lim_{x \to z} v(x)\right] = A \cdot B$$

3) $\lim_{x \to a} \dfrac{u(x)}{v(x)} = \dfrac{\lim\limits_{x \to a} u(x)}{\lim\limits_{x \to a} v(x)} = \dfrac{A}{B}$, $(B \neq 0)$

4) $\lim_{x \to a} [u(x)]^n = \left[\lim_{x \to a} u(x) \right]^n = A^n$

5) $\lim_{x \to a} u(x) = \lim_{x \to a} v(x)$, If $u(x) = v(x)$ $(x \neq a)$

6) $\lim_{x \to \infty} \dfrac{1}{x^n} = 0$, and $\lim_{x \to -\infty} \dfrac{1}{x^n}$, n is a positive integer

where
> a, k, h, n, A, and B are real numbers.

3. Slope of Tangent Line

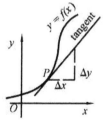

The gradient m of a curve $y = f(x)$ varies from point to point. The gradient of a curve is the slope of the tangent at some point P of a curve $y = f(x)$:

$$m = \frac{\Delta y}{\Delta x}$$

---

4. Definition of the Derivative

For any function, $y = f(x)$ between points P and P_1,

$$\frac{\Delta y}{\Delta x} = \frac{f(x) + \Delta x - f(x)}{\Delta x},$$

is the average rate of change of the function $y = f(x)$, and it is the derivative of the function $y = f(x)$. The process of finding this limit, the derivative, is called *differentiation.*

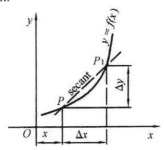

The derivative of the function may be denoted in any of the following ways,

$$f'(x), \quad y', \quad \frac{dy}{dx}, \text{ or } \frac{d}{dx}\big[f(x)\big]$$

Hence,

$$y' = \lim_{\Delta x \to 0} \frac{\Delta y}{\Delta x} = \lim_{\Delta x \to 0} \frac{f(x + \Delta x) - f(x)}{\Delta x}$$

5. Basic Derivatives

Function	Derivative
Basic rules	
$y = k,$ k is a real number	$y' = 0$
$y = c \cdot x^n + C$	$y' = c \cdot n \cdot x^{n-1}$
$y = u(x) \pm v(x)$	$y' = u'(x) \pm v'(x)$
$y = u(x) \cdot v(x)$	$y' = u' \cdot v + u \cdot v'$
$y = \dfrac{u(x)}{v(x)}$ $v(x) \neq 0$	$y' = \dfrac{u' \cdot v - u \cdot v'}{v^2}$
$y = \sqrt{x}$	$y' = \dfrac{1}{2\sqrt{x}}$
Chain Rule	
$y = f[u(x)]$	$y' = f'(u) \cdot u'(x) = $ $= \dfrac{dy}{dx} = \dfrac{dy}{du} \cdot \dfrac{du}{dx}$
Parametric Form of Derivative	
$y = f(x)$ $\begin{cases} x = f(t), \\ y = f(t) \end{cases}$	$y' = \dfrac{dy}{dt} \cdot \dfrac{dt}{dx};\ y'' = \dfrac{d^2 y}{dx^2}$
Derivative of Exponential Functions	
$y = e^x$	$y' = e^x = y''$

Derivatives

Continued from # 5

$y = e^{-x}$	$y' = -e^{-x}$
$y = e^{ax}$	$y' = a \cdot e^{ax}$
$y = x \cdot e^x$	$y' = e^x(1 + x)$
$y = \sqrt{e^x}$	$y' = \dfrac{\sqrt{e^x}}{2}$
$y = a^x$	$y' = a^x \ln a$
$y = a^{nx}$	$y' = n \cdot a^{nx} \ln a$
$y = a^{x^2}$	$y' = a^{x^2} \cdot 2x \ln a$
Derivative of Trigonometric Functions	
$y = \sin x \quad y' = \cos x$ $y = \cos x$	$y' = \cos x$ $y' = -\sin x$
$y = \tan x$	$y' = \dfrac{1}{\cos^2 x} = 1 + \tan^2 x$
$y = \cot x$	$y' = \dfrac{-1}{\sin^2 x} = -(1 + \cot^2 x)$
$y = a \cdot \sin(kx)$	$y' = a \cdot k \cdot \cos(kx)$
$y = a \cdot \cos(kx)$	$y' = -a \cdot k \cdot \sin(kx)$
$y = \sin^n x$	$y' = n \cdot \sin^{n-1} x \cdot \cos x$

Continued from # 5

$y = \cos^n x$	$y' = -n\cos^{n-1} x \sin x$
$y = \tan^n x$	$y' = n\tan^{n-1} x\left(1 + \tan^2 x\right)$
$y = \cot^n x$	$y' = -n \cdot \cot^{n-1} x \cdot \left(1 + \cot^2 x\right)$
$y = \dfrac{1}{\sin x}$	$y' = \dfrac{-\cos x}{\sin^2 x}$
$y = \dfrac{1}{\cos x}$	$y' = \dfrac{\sin x}{\cos^2 x}$

Derivative of Inverse Trigonometric Functions

$y = \arcsin x$	$y' = \dfrac{1}{\sqrt{1 - x^2}}$
$y = \arccos x$	$y' = -\dfrac{1}{\sqrt{1 - x^2}}$
$y = \arctan x$	$y' = \dfrac{1}{1 + x^2}$
$y = \operatorname{arc} \cot x$	$y' = -\dfrac{1}{1 + x^2}$
$y = \operatorname{arcsinh} x$	$y' = \dfrac{1}{\sqrt{x^2 + 1}}$

Continued from # 5

$y = \text{arccosh } x$	$y' = \dfrac{1}{\sqrt{x^2 - 1}}$
$y = \text{arctanh } x$	$y' = \dfrac{1}{1 - x^2}$
$y = \text{arccoth } x$	$y' = \dfrac{1}{1 - x^2}$
Derivative of Hyperbolic Functions	
$y = \sinh x$	$y' = \cosh x$
$y = \cosh x$	$y' = \sinh x$
$y = \tanh x$	$y' = \dfrac{1}{\cosh^2 x}$
$y = \coth x$	$y' = -\dfrac{1}{\sinh^2 x}$
Derivative of Logarithmic Functions	
$y = \ln x$	$y' = \dfrac{1}{x},$
$y = \log_a x$	$y' = \dfrac{1}{x \cdot \ln a}$
$y = \ln(1 \pm x)$	$y' = \pm \dfrac{1}{1 \pm x}$

Continued from 5

$y = \ln x^n$	$y' = \dfrac{n}{x}$
$y = \ln \sqrt{x}$	$y' = \dfrac{1}{2x}$

6. Increasing and Decreasing Function $y = f(x)$

If $y'(x) > 0$, function $y(x)$ increases for each value of x an interval (a, b).

If $y'(x) < 0$, function $y(x)$ decreases for each value of x an interval (a, b).

If $y'(x) = 0$, function $y(x)$ is tangentially parallel to the x-axis at x.

7. Maximum and Minimum Function $y = f(x)$

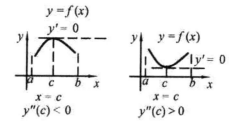

If $y''(c) > 0$, and $y'(c) = 0$, there is a minimum at $x = c$.

If $y''(c) < 0$, and $y'(c) = 0$, there is a maximum at $x = c$.

If $y''(c) = 0$, then the test gives no information.

8. Solving Applied Problems

Step 1: Read problem carefully.

Step 2: If possible, sketch a diagram.

Step 3: Decide on the variable whose values must be maximized or minimized. Express that variable as a function of one other variable.

Step 4: Find the critical points for the function of Step 3. Check these for maximum or minimum.

Step 5: Check the extrema at any end point of the domain of the function of Step 3.

Step 6: Check to be sure the answer is reasonable.

9. Integration

Integration is the opposite of derivation In calculus integration of a given real valid function $y = f(x)$ is a function $F(x)$ whose derivative is equal to $f(x)$, i.e.,

$$F'(x) = \frac{dF(x)}{dx} = f(x)$$

There are two meanings of integration: definite integrals and indefinite integrals.

---

a) Indefinite integrals

The integral of a function is a special limit with many diverse applications.

If $F'(x) = f(x)$, then

$$\int f(x)dx = F(x) + C$$

where
C = unknown constant

b) Definite integral

If $f(x)$ is continuous on the interval $[a, b]$, the definite integral of $f(x)$ from a to b is given by

$$\int_a^b f(x)dx = F(x)\Big|_a^b = F(b) - F(a)$$

10. Basic Integration Rules

1) The indefinite integral of a constant
$$\int kdx = kx + C \quad (k = \text{constant})$$

2) The power rule for indefinite integrals
$$\int x^n dx = \frac{1}{n+1} x^{n+1} + C$$

3) The indefinite integral of a constant multiple of a function

$$\int c \cdot f(x)dx = c \int f(x)dx \quad (c = \text{constant})$$

4) The sum rule

$$\int [f(x) \pm g(x)]dx = \int f(x)dx \pm \int g(x)dx$$

5) The indefinite integral of the exponential function

$$\int e^x dx + C$$

6) The indefinite integral of the function $f(x) = x^{-1}$

$$\int x^{-1} dx = \int \frac{1}{x} dx = \ln|x| + C \quad (x \neq 0)$$

11. Integration by Substitution

The method of substitution is related to the chain rule for differentiating functions.

There are five steps involved in integration by substitution.

Consider the indefinite integral

$$\int f[g(x)]g'(x)dx$$

Step 1: Let $u = g(x)$, where $g(x)$ is part of the integrand, usually the inside function of the composite function $f[g(x)]$.

Step 2: Determine $g'(x) = \dfrac{du}{dx}$

Step 3: Use the substitute $u = g(x)$ and $du = g'(x)dx$
to convert the entire integral into one involving
only u.

Step 4: Evaluating the resulting integral.

Step 5: Replace u with $g(x)$ to obtain the final solution as
a function of x.

Example:
Find

$$F(x) = \int \sqrt{3x - 5}\, dx$$

Solution:

Step 1: Observe that the integrand involves the
composite function $\sqrt{3x - 5}$ with the "inside
function" $g(x) = 3x - 5$. So, we choose
$$u = 3x - 5$$

Step 2: Compute $du = g'(x) = 3dx$

Step 3: Making the substitution $u = 3x - 5$ and
$du = g'(x) = 3dx$, we obtain

$$F(x) = \frac{1}{3} \int \sqrt{u}\, du$$

Step 4: Evaluate

$$F(x) = \frac{1}{3} \int \sqrt{u}\, du = \frac{2}{9} u\sqrt{u} + C$$

Step 5: Replacing u with $3x-5$ we obtain

$$F(x) = \frac{2}{9}(3x-5)\sqrt{3x-5} + C$$

12. Basic Integrals

1) $\int \dfrac{1}{x^n dx} = -\dfrac{1}{n-1} \cdot \dfrac{1}{x^{n-1}} + C \qquad (n \neq 1)$

2) $\int a^{bx}\, dx = \dfrac{1}{b} \cdot \dfrac{a^{bx}}{\ln|a|} + C$

3) $\int (\ln x)^2\, dx = x(\ln|x|)^2 - 2x\ln|x| + 2x + C$

4) $\int \dfrac{dx}{\ln x} = \ln\left|(\ln|x|)\right| + \ln|x| + \dfrac{(\ln|x|)^2}{2 \cdot 2!} + \dfrac{(\ln|x|)^3}{3 \cdot 3!} + \cdots$

5) $\int x\ln x\, dx = x^2\left[\dfrac{\ln|x|}{2} - \dfrac{1}{4}\right] + C$

6) $\int \dfrac{dx}{x\ln x} = \ln\left|(\ln|x|)\right| + C$

7) $\int e^{ax}\, dx = \dfrac{1}{a} e^{ax} + C$

8) $\int x e^{ax}\, dx = \dfrac{e^{ax}}{a^2}(ax-1) + C$

9) $\int x^2 e^{ax} dx = e^{ax}\left(\dfrac{x^2}{a} - \dfrac{2x}{a^2} + \dfrac{2}{a^3}\right) + C$

10) $\int x^n e^{ax} dx = \dfrac{1}{a} x^n e^{ax} - \dfrac{n}{a}\int x^{n-1} e^{ax} dx + C$

11) $\int \dfrac{dx}{1 + e^{ax}} = \dfrac{1}{a}\ln\left|\dfrac{e^{ax}}{1 + e^{ax}}\right| + C$

12) $\int \dfrac{e^{ax} dx}{b + ce^{ax}} = \dfrac{1}{ac}\ln\left|b + ce^{ax}\right| + C$

13) $\int e^{ax\ln x} dx = \dfrac{e^{ax}\ln|x|}{a} - \dfrac{1}{a}\int e^{ax} x\, dx + C$

14) $\int e^{ax}\cos bx\, dx = \dfrac{e^{ax}}{a^2 + b^2}\left(a\cos bx + b\sin bx\right)$

15) $\int e^{ax}\sin bx\, dx = \dfrac{e^{ax}}{a^2 + b^2} + \left(a\sin bx - b\cos bx\right)$

16) $\dfrac{dx}{ax + b} = \dfrac{1}{a}\ln\left|ax + b\right| + C$

17) $\int \dfrac{dx}{(ax + b)^n} = \dfrac{1}{a(n-1)(ax + b)^{n-1}} + C \quad (n \neq 1)$

18) $\int \dfrac{dx}{ax - b} = \dfrac{1}{a}\ln\left|ax - b\right| + C$

19) $\int \dfrac{dx}{(ax - b)^n} = \dfrac{1}{a(n-1)(ax - b)^{n-1}} + C \quad (n \neq 1)$

---

20) $\int \dfrac{dx}{(ax+b)(cx+d)} = \dfrac{1}{bc-ad} \ln\left|\dfrac{cx+d}{ax+b}\right| + C \left(bc\text{-}ad \neq 0\right)$

21) $\int \dfrac{dx}{(ax-b)(cx-d)} = \dfrac{1}{bc-ad} \ln\left|\dfrac{cx-d}{ax-b}\right| + C,$

$$\left(bc\text{-}ad \neq 0\right)$$

22) $\int \dfrac{x\,dx}{ax+b} = \dfrac{x}{a} - \dfrac{b}{a^2} \ln|ax-b| + C$

23) $\int \dfrac{x^2\,dx}{ax+b} = \dfrac{1}{a^3}\left[\begin{array}{l} \dfrac{1}{2}(ax+b)^2 - 2b(ax+b) + \\ + b^2 \ln|ax+b| \end{array}\right] + C$

24) $\int \dfrac{dx}{x(ax+b)} = -\dfrac{1}{b}\ln\left|a+\dfrac{b}{x}\right| + C$

26) $\int \dfrac{x^3\,dx}{ax+b} = \dfrac{1}{a^4}\left[\begin{array}{l} \dfrac{1}{3}(ax+b)^3 - \dfrac{3}{2}b(ax+b)^2 + \\ + 3b^2(ax+b) - b^3 \ln|ax+b| \end{array}\right] + C$

27) $\int \dfrac{dx}{a^2+x^2} = \dfrac{1}{a}\arctan\dfrac{x}{a} + C$

28) $\int \dfrac{x\,dx}{a^2+x^2} = \dfrac{1}{2}\ln\left|a^2+x^2\right| + C$

29) $\int \dfrac{x^2\,dx}{a^2+x^2} = x - a\arctan\dfrac{x}{a} + C$

30) $\int \dfrac{x^3\,dx}{a^2+x^2} = \dfrac{x^2}{2} - \dfrac{a^2}{2}\ln\left|a^2+x^2\right| + C$

31) $\int \dfrac{dx}{a^2 - x^2} = -\int \dfrac{dx}{x^2 - a^2} = \dfrac{1}{2a} \ln \left| \dfrac{a+x}{a-x} \right| + C$

32) $\int \dfrac{xdx}{a^2 - x^2} = -\int \dfrac{xdx}{x^2 - a^2} = -\dfrac{1}{2} \ln \left| a^2 - x^2 \right| + C$

33) $\int \dfrac{x^2 \, dx}{a^2 - x^2} = -\int \dfrac{x^2 \, dx}{x^2 - a^2} = -x + a\dfrac{1}{2} \ln \left| \dfrac{a+x}{a-x} \right| + C$

34) $\int \dfrac{x^3 \, dx}{a^2 - x^2} = -\int \dfrac{x^3 \, dx}{x^2 - a^2} = -\dfrac{x^2}{2} - \dfrac{a^2}{2} \ln \left| a^2 - x^2 \right| + C$

35) $\int \dfrac{xdx}{\left(a^2 + x^2 \right)^2} = -\dfrac{1}{2\left(a^2 + x^2 \right)} + C$

36) $\int \dfrac{x^2 \, dx}{\left(a^2 + x^2 \right)^2} = -\dfrac{x}{2\left(a^2 + x^2 \right)} + \dfrac{1}{2a} \arctan \dfrac{x}{a} + C$

37) $\int \dfrac{x^3 \, dx}{\left(a^2 + x^2 \right)^2} = \dfrac{a^2}{2\left(a^2 + x^2 \right)} + \dfrac{1}{2} \ln \left| a^2 + x^2 \right| + C$

38) $\int \dfrac{dx}{\left(a^2 - x^2 \right)^2} = \dfrac{x}{2a^2 \left(a^2 - x^2 \right)} + \dfrac{1}{2a^3} \cdot \dfrac{1}{2} \ln \left| \dfrac{2+x}{a-x} \right| + C$

39) $\int \dfrac{xdx}{\left(a^2 - x^2 \right)^2} = \dfrac{1}{2\left(a^2 - x^2 \right)} + C$

40) $\int \dfrac{x^2 \, dx}{\left(a^2 - x^2 \right)^2} = \dfrac{x}{2\left(a^2 - x^2 \right)} - \dfrac{1}{2a} \cdot \dfrac{1}{2} \ln \left| \dfrac{2+x}{a-x} \right| + C$

Basic Integrals

41) $\int \dfrac{x^3\,dx}{\left(a^2-x^2\right)^2} = \dfrac{a^2}{2\left(a^2-x^2\right)} + \dfrac{1}{2}\ln\left|a^2-x^2\right| + C$

42) $\int \sqrt{x}\,dx = \dfrac{2}{3}\sqrt{3} + C$

43) $\int \sqrt{ax+b}\,dx = \dfrac{2}{3a}\sqrt{(ax+b)^3} + C$

44) $\int x\sqrt{ax+b}\,dx = \dfrac{2(3ax-2b)\sqrt{(ax+b)^3}}{15a^2} + C$

45) $\int \dfrac{dx}{\sqrt{x}} = 2\sqrt{x} + C$

46) $\int \dfrac{dx}{\sqrt{ax+b}} = \dfrac{2\sqrt{(ax+b)}}{a} + C$

47) $\int \dfrac{x\,dx}{\sqrt{ax+b}} = \dfrac{2(ax-ab)\sqrt{(ax+b)}}{3a^2} + C$

48) $\int \dfrac{x^2\,dx}{\sqrt{ax+b}} = \dfrac{2\left(2a^2x^2 - 4abx + 8b^2\right)\sqrt{(ax+b)}}{15a^3} + C$

49) $\int \sqrt{a^2+x^2}\,dx = \dfrac{x}{2}\sqrt{a^2+x^2} + \dfrac{a^2}{2}\operatorname{arcsin} h\dfrac{x}{a} + C$

50) $\int x\sqrt{a^2+x^2}\,dx = \dfrac{1}{3}\sqrt{(a^2+x^2)^3} + C$

51) $\int x^3 \sqrt{a^2 + x^2}\, dx = \dfrac{\sqrt{(a^2 + x^2)^5}}{5} - \dfrac{a^2 \sqrt{(a^2 + x^2)^3}}{3} + C$

52) $\int x^2 \sqrt{a^2 + x^2}\, dx = \dfrac{x}{4} \sqrt{(a^2 + x^2)^3} -$

$\qquad - \dfrac{a^2}{8}\left(x\sqrt{a^2 + x^2}\right) + a^2 \operatorname{arcsin} h \dfrac{x}{a} + C$

53) $\int \dfrac{\sqrt{a^2 + x^2}}{x}\, dx = \sqrt{a^2 + x^2} - a\ln\left|\dfrac{a + \sqrt{a^2 + x^2}}{x}\right| + C$

54) $\int \dfrac{\sqrt{a^2 + x^2}}{x^2}\, dx = -\dfrac{\sqrt{a^2 + x^2}}{x} + \operatorname{arcsin} h \dfrac{x}{a} + C$

55) $\int \dfrac{dx}{\sqrt{a^2 + x^2}} = \operatorname{arcsin} h \dfrac{x}{a} + C$

56) $\int \dfrac{x\, dx}{\sqrt{a^2 + x^2}} = \sqrt{a^2 + x^2} + C$

57) $\int \dfrac{x^2\, dx}{\sqrt{a^2 + x^2}} = \dfrac{x}{2} \sqrt{a^2 + x^2} - \dfrac{a^2}{2} \operatorname{arcsin} h \dfrac{x}{a} + C$

58) $\int \dfrac{x^3\, dx}{\sqrt{a^2 + x^2}} = \dfrac{\sqrt{(a^2 + x^2)^3}}{3} - a^2 \sqrt{x^2 + a^2} + C$

59) $\int \dfrac{dx}{x\sqrt{a^2 + x^2}} = -\dfrac{1}{a}\ln\left|\dfrac{a + \sqrt{a^2 + x^2}}{x}\right| + C$

60) $\int \dfrac{dx}{x^2 \sqrt{a^2 + x^2}} = -\dfrac{\sqrt{x^2 + a^2}}{a^2 x} + C$

61) $\int x\sqrt{a^2 - x^2}\, dx = -\dfrac{1}{3}\sqrt{\left(a^2 - x^2\right)^3} + C$

62) $\int \dfrac{dx}{\sqrt{a^2 - x^2}} = \arcsin\dfrac{x}{a} + C$

63) $\int \dfrac{x\,dx}{\sqrt{a^2 - x^2}} = -\sqrt{a^2 - x^2} + C$

64) $\int \dfrac{x^2\,dx}{\sqrt{a^2 - x^2}} = -\dfrac{x}{2}\sqrt{a^2 - x^2} + \dfrac{a^2}{2}\arcsin\dfrac{x}{a} + C$

65) $\int \dfrac{x^3\,dx}{\sqrt{a^2 - x^2}} = \dfrac{\sqrt{\left(a^2 - x^2\right)^3}}{3} - a^2\sqrt{x^2 - a^2} + C$

66) $\int \dfrac{dx}{x\sqrt{a^2 - x^2}} = -\dfrac{1}{a}\ln\left|\dfrac{a + \sqrt{a^2 - x^2}}{x}\right| + C$

67) $\int \dfrac{dx}{x^2 \sqrt{a^2 - x^2}} = -\dfrac{\sqrt{x^2 - a^2}}{a^2 x} + C$

68) $\int \dfrac{\sqrt{a^2 - x^2}}{x}\, dx = \sqrt{a^2 - x^2} - a\arccos\dfrac{a}{x} + C$

69) $\int \dfrac{\sqrt{a^2 - x^2}}{x^2}\, dx = -\dfrac{\sqrt{x^2 - a^2}}{x}\arccos h\dfrac{a}{x} + C$

70) $\int \dfrac{\sqrt{a^2 - x^2}}{x^3} dx = -\dfrac{\sqrt{x^2 - a^2}}{2x} + \dfrac{1}{2a} \arccos \dfrac{a}{x} + C$

71) $\int \cos x dx = \sin x + C$

72) $\int \sin x dx = -\cos x + C$

73) $\int \tan x dx = \ln|\sec x| + C$

74) $\int \cot x dx = \ln|\sin x| + C$

75) $\int \sec x dx = \ln|\sec x + \tan x| + C$

76) $\int \csc x dx = \ln|\csc x - \cot x| + C$

77) $\int \csc^2 x dx = -\cot x + C$

78) $\int \sec x \tan x dx = \sec x + C$

79) $\int \csc x \cot x dx = -\csc x + C$

80) $\int \sin(bx) dx = -\dfrac{1}{b} \cos(bx) + C$

81) $\int \sin^2(bx) dx = \dfrac{x}{2} - \dfrac{1}{4b} \sin(2bx) + C$

82) $\int \cos(bx) dx = \dfrac{1}{b} \sin(bx) + C$

83) $\int \tan(bx) dx = \dfrac{1}{b} \ln|\sec(bx)| + C$

84) $\int \sec(bx) dx = \dfrac{1}{b} \ln|\tan(bx) + \sec(bx)| + C$

85) $\int \sin(ax)\sin(bx)dx = \dfrac{\sin(a-b)x}{2(a-b)} - \dfrac{\sin(a+b)x}{2(a+b)} + C$

$$(|a| \neq |b|).$$

86) $\int \cos(ax)\cos(bx)dx = \dfrac{\sin(a-b)x}{2(a-b)} + \dfrac{\sin(a+b)x}{2(a+b)} + C$

$$(|a| \neq |b|)$$

87) $\int \sin(ax)\cos(bx)dx = \dfrac{\cos(a-b)x}{2(a-b)} - \dfrac{\cos(a+b)x}{2(a+b)} + C$

$$(|a| \neq |b|)$$

88) $\int x^n \sin bx\, dx = -\dfrac{x^n}{b}\cos bx + \dfrac{n}{b}\int x^{n-1}\cos bx\, dx + C$

89) $\int x^n \cos bx\, dx = \dfrac{x^n}{b}\sin bx - \dfrac{n}{b}\int x^{n-1}\sin bx\, dx + C$

90) $\int \sin^n x\, dx = -\dfrac{1}{n}\sin^{n-1} x\cos x + \dfrac{n-1}{n}\int \sin^{n-2} x\, dx + C$

91) $\int \cos^n x\, dx = \dfrac{1}{n}\cos^{n-1} x\sin x + \dfrac{n-1}{n}\int \cos^{n-2} x\, dx + C$

92) $\int \arcsin x\, dx = x\arcsin x + \sqrt{1-x^2} + C$

93) $\int \arccos x\, dx = x\arccos x - \sqrt{1-x^2} + C$

---

94) $\int \arctan x\, dx = x \arctan x - \dfrac{1}{2} \ln\left|1 + x^2\right| + C$

95) $\int \text{arc cot } x\, dx = x \text{ arc cot } x + \dfrac{1}{2} \ln\left|1 + x^2\right| + C$

96) $\int \sinh(ax)\, dh = \dfrac{1}{a} \cosh(ax) + C$

97) $\int \sinh^2 x\, dx = \dfrac{1}{4} \sinh(2x) - \dfrac{x}{2} + C$

13. Arc Length

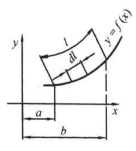

a) Arc differential:

$$dl = \sqrt{dx^2 + dy^2} = \sqrt{1 + \left(\dfrac{dy}{dx}\right)^2}\, dx$$

b) Arc length

Length of curve $y = f(x)$ from $x = a$ to $x = b$ is

$$l = \int_a^b \sqrt{1 + y'^2}\, dx; \quad y' = \frac{dy}{dx}$$

c) The surface area

Surface area where the curve $y = f(x)$ rotates around the x-axis is

$$A = 2\pi \int_a^b y\sqrt{1 + y'^2}\, dx$$

14. Finding an Area and a Volume

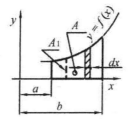

a) Area

Area A below the curve $y = f(x)$ from $x = a$ to $x = b$ is

$$A = \int_a^b y\, dx$$

___

b) Volume

1) Volume of a rotating body where area A rotates around the x-axis:

$$V = \pi \int_a^b y^2\, dx$$

2) Volume of a body the cross section $A1$ of which is a function of x:

$$V = \int_a^b A_1\, dx$$

15. Finding the Area between Two Curves

Area A between curve $y = f(x)$ and $y = g(x)$ from $x = a$ to $x = b$ is

$$A = \int_a^b f(x)\,dx - \int_a^b g(x)\,dx = \int_a^b \left[f(x) - g(x) \right] dx$$

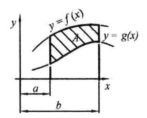

Example:

Find area of the region A bounded by the graphs of $f(x) = 2x - 1$ and $g(x) = x^2 - 3$ and the vertical lines $x = 0$ and $x = 2$.

Solution:

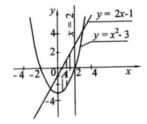

$$\int_0^2 [f(x) - g(x)]dx = \int_0^2 [(2x - 1) - (x^2 - 3)]dx$$

$$= \int_0^2 (-x^2 + 2x + 2)dx$$

$$= -\frac{1}{3}x^3 + x^2 + 2x \Big|_0^2$$

$$= \left(-\frac{8}{3} + 4 + 4\right) - \left(-\frac{1}{3} + 1\right) = \frac{14}{3}$$

Hence, area $A = \dfrac{14}{3} = 4\dfrac{2}{3}$

STATISTICS

Statistics is the mathematics of the collection, organization, and interpretation of numerical data, especially the analysis of population characteristics by inference from sampling. The most familiar statistical measure is the arithmetic mean, which is an average value for a group of numerical observations.

A second important statistic or statistical measure is the standard deviation, which is a measure of how much the individual observations are scattered about the mean.

This section contains the most frequently used formulas, rules, and definitions regarding to the following:

1. Sets
2. Permutations and Combinations
3. Probability
4. Distribution
5. Reliability

---

1. Definition of Set and Notation

A set is a collection of object called elements. In mathematics we write a set by putting its elements between the curly brackets $\{\ \}$.

Set A which containing numbers 3, 4, and 5 is written

$$A = \{3, 4, 5\}$$

a) Empty set

A set with no elements is called an empty set and is denoted by

$$\{\ \} = \Phi$$

b) Subset

Sometimes every element of one set also belongs to another set:

$$A = \{3, 4, 5\} \text{ and } B = \{1, 2, 3, 4, 5, 6, 7\},$$

A set A is a subset of a set B because every element of set A is also an element of set B, and it is written as

$$A \subseteq B$$

c) Set equality

The sets A and B are equal if and only if they have exactly the same elements, and the equality is written as

$$A = B$$

---

d) Set union

The union of a set A and set B is the set of all elements that belong to either A or B or both, and is written as

$$A \cup b = \{x \,|\, x \in A \ \text{or} \ x \in B \ \text{or} \ \text{both}\}$$

2. Terms and Symbols

$\{\ \}$ set braces

$\in$ is an element of

$\notin$ is not an element of

$\subseteq$ is a subset of

$\not\subset$ is not a subset of

A' complement of set A

$\cap$ set intersection

$\cup$ set union

3. Venn Diagrams

Venn diagrams are used to visually illustrate relationships between sets.

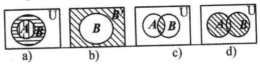

These Venn diagrams illustrate the following statements:

a) Set A is a subset of set B ($A \subset B$).
b) Set B' is the complement of B.
c) Two sets A and B with their intersection $A \cap B$.
d) Two sets A and B with their union $A \cup B$.

4. Operations on Sets

If A, B and C are arbitrary subsets of universal set U, then the following rules govern the operations on sets:

1) Commutative law for union

$$A \cup B = B \cup A$$

2) Commutative law for intersection

$$A \cap B = B \cap A$$

3) Associative law for union

$$A \cup (B \cup C) = (A \cup B) \cup C$$

4) Associative law for intersection

$$A \cap (B \cap C) = (A \cap B) \cap C$$

5) Distributive law for union

$$A \cup (B \cap C) = (A \cup B) \cap (A \cup C)$$

6) Distributive law for intersection

$$A \cap (B \cap C) = (A \cap B) \cup (A \cap C)$$

5. De Morgan's Laws

$$\left(A \cup B\right)' = A' \cap B' \quad (1)$$

$$\left(A \cap B\right)' = A' \cup B' \quad (2)$$

The complement of the union of two sets is equal to the intersection of their complements (equation 1).
The complement of the intersection of two sets is equal to the union of their complements (equation 2).

6. Counting the Elements in a Set

The number of the elements in a finite set is determined by simply counting the elements in the set.

If A and B are disjoint sets, then

$$n(A \cup B) = n(A) + n(B)$$

In general, A and B need not to be disjoint, so

$$n(A \cup B) = n(A) + n(B) - n(A \cap B)$$

where

n = number of the elements in a set

7. Permutations

A permutation of m elements from a set of n elements is any arrangement, without repetition, of the m elements. The total number of all the possible permutations of n distinct objects taken m times is

$$P(n, m) = \frac{n!}{(n-m)!}, \qquad (n \geq m)$$

Example:

Find the number of ways a president, vice-president, secretary, and a treasurer can be chosen from a committee of eight members.

Solution:

$$P(n, m) = \frac{n!}{(n-m)!} = P(8,4) = \frac{8!}{(8-4)!} =$$

$$= \frac{8 \cdot 7 \cdot 6 \cdot 5 \cdot 4 \cdot 3 \cdot 2 \cdot 1}{4 \cdot 3 \cdot 2 \cdot 1} = 1680$$

There are 1,680 ways of choosing the four officials from the committee of eight members.

---

8. Combinations

The number of combination of n distinct elements taken is given by

$$C(n, m) = \frac{n!}{m!(n-m)!}, \quad (n \geq m)$$

Example:
How many poker hands of five cards can be dealt from a standard deck of 52 cards?

Solution:
Note: The order in which the 5 carts are dealt is not important.

$$C(n, m) = \frac{n!}{m!(n-m)!} = C(52, 5) = \frac{52!}{5!(52-5)!} = \frac{52!}{5! \, 47!}$$

$$= \frac{52 \cdot 51 \cdot 50 \cdot 49 \cdot 48}{5 \cdot 4 \cdot 3 \cdot 2 \cdot 1} = 2{,}598{,}963$$

9. Probability Terminology

A number of specialized terms are used in the study of probability.

Experiment: An experiment is an activity or occurrence with an observable result.

Outcome: The result of the experiment.

Sample point: An outcome of an experiment.
Event: An event is a set of outcomes (a subset of the sample space) to which a probability is assigned.

10. Basic Probability Principles

Consider a random sampling process in which all the outcomes solely depend on chance, i.e., each outcome is equally likely to happen. If S is a uniform sample space and the collection of desired outcomes is E, the probability of the desired outcomes is

$$P(E) = \frac{n(E)}{n(S)}$$

where

$n(E)$ = number of favorable outcomes in E
$n(S)$ = number of possible outcomes in S

Since E is a subset of S,

$$0 \le n(E) \le n(S),$$

the probability of the desired outcome is

$$0 \le P(E) \le 1$$

11. Random Variable

A random variable is a rule that assigns a number to each outcome of a chance experiment.

Example:

1. A coin is tossed six times. The random variable X is the number of tails that are noted. X can only take the values 1, 2,…, 6, so X is a discrete random variable.

2. A light bulb is burned until it burns out. The random variable Y is its lifetime in hours. Y can take any positive real value, so Y is a continuous random variable.

12. Mean Value $\bar{x}$ or Expected Value μ

The mean value or expected value of a random variable indicates its average or central value. It is a useful summary value of the variable's distribution.

1) If random variable X is a discrete mean value,

$$\bar{x} = x_1 p_1 + x_2 p_2 + \ldots + x_n p_n = \sum_{i=1}^{n} x_i p_i$$

where

p_i = probability densities

2) If X is a continuous random variable with probability density function $f(x)$, then the expected value of X is

$$\mu = E(X) = \int_{-\infty}^{+\infty} xf(x)dx$$

where

$f(x) =$ probability densities

13. Variance

The variance is a measure of the "spread" of a distribution about its average value.

a) Discrete system:

$$\sigma^2 = \sum_{i=1}^{n} \left(x_i - \bar{x} \right)^2 p_i$$

b) Continuous system:

$$\sigma^2 = \int_{-\infty}^{+\infty} (x - \mu)^2 \cdot f(x)dx$$

14. Standard Deviation

Standard deviation, denoted by σ, is the positive square root of the variance. Both variance and standard deviation are used to describe the spread of a distribution.

a) Discrete system:

$$\sigma = \sqrt{\sum_{i=1}^{n} \left(x_i - \bar{x} \right)^2 p_i}$$

b) Continuous system:

$$\sigma = \sqrt{\int_{-\infty}^{+\infty} (x - \mu)^2 \cdot f(x)dx}$$

15. Normal Distribution

The normal distribution, or Gaussian distribution, is a symmetrical distribution commonly referred to as a bell curve.

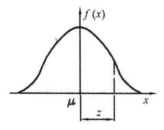

a) Probability density function:

$$f(x) = \frac{1}{\sigma\sqrt{2\pi}} e^{\frac{-(x-\mu)^2}{2\sigma^2}}$$

b) Distribution Function:

$$F(x) = \int_{-\infty}^{x} \frac{1}{\sigma\sqrt{2\pi}} e^{\frac{-(t-\mu)^2}{2\sigma^2}} dt$$

c) Standard value (z-score)

If normal distribution has mean μ and standard deviation σ, then the z-score for the number x is

$$z = \frac{x - \mu}{\sigma}.$$

16. Binomial Distribution

Binomial distribution, also known as Bernoulli distribution, describes the random sampling processes such that all outcomes are either yes or no (success/failure) without ambiguity.

Suppose that the probability of success in a single trial is p in a random sampling process and the failure rate is q where,

$$q = 1 - p,$$

the binomial distribution with exactly x successes in n trials, where $x \le n$, has the following important properties.

a) Density function:

$$f(x) = \frac{n!}{x(n-x)!} \, p^n q^{n-x}$$

b) Mean:

$$\mu = np$$

c) Variance:

$$\sigma^2 = npq$$

d) Standard deviation:

$$\sigma = \sqrt{npq}$$

17. Poisson Distribution

The Poisson distribution describes a random sampling process in which the desired outcomes occur relatively infrequently but at a regular rate.

Suppose there are on average λ successes in a large number of trials (large sampling period). The Poisson distribution with exactly x successes in the same sampling period has the following important properties.

a) Density function:

$$f(x) = \frac{\lambda^x e^{-\lambda}}{x!}$$

b) Mean:

$$\mu = \lambda = np$$

c) Distribution function:

$$F(x_j) = \sum_{k \leq j} \frac{\lambda^{x_k} e^{-\lambda}}{x_k}$$

d) Variance:

$$\sigma^2 = \lambda = np$$

e) Standard deviation:

$$\sigma = \sqrt{\lambda} = \sqrt{np} \ , \quad (\lambda = \text{constant} > 0)$$

18. Exponential Distribution

The exponential distribution is used for reliability calculation.

a) Density function:

$$f(x) = \lambda e^{-\lambda x}, \quad (\lambda > 0, \ x \geq 0)$$

b) Distribution function:

$$F(x) = 1 - e^{-\lambda x}$$

c) Mean:

$$\mu = \frac{1}{\lambda}$$

d) Variance:

$$\sigma^2 = \frac{1}{\lambda^2}$$

e) Standard deviation:

$$\sigma = \sqrt{\frac{1}{\lambda^2}} = \frac{1}{\lambda}$$

19. General Reliability Definitions

a) Reliability function

The reliability function $R(t)$, also known as the survival function $S(t)$, is defined by

$$R(t) = S(t) = 1 - F(t)$$

b) Failure distribution function

The failure distribution function is the probability of an item failing in the time interval $0 \leq \tau \leq t$

$$F(t) = \int_0^t f(\tau)d\tau, \quad (t \geq 0)$$

c) Failure rate

The failure rate of the unit is

$$z(t) = \lim_{\Delta t \to 0} \frac{F(t - \Delta t)}{R(t)} = \frac{f(t)}{R(t)}$$

---

d) Mean time to failure

The mean time to failure (MTTF) of a unit is

$$MTTF = \int_{0}^{\infty} f(t) \cdot t \, dt = \int_{0}^{\infty} R(t) \, dt$$

e) Reliability of the system

The reliability of the system is the product of the reliability functions of the components $R_1, \cdots, R_n$

$$R_S(t) = R_1 \cdot R_2 \cdot \ldots \cdot R_n = \prod_{i=1}^{n} R_i(t)$$

20. Exponential Distribution Used as Reliability Function

a) Reliability function:

$$R(t) = e^{-\lambda t} \qquad (\lambda = \text{constant})$$

b) Failure distribution function:

$$F(t) = 1 - e^{-\lambda t}$$

c) Density function of failure:

$$f(t) = \lambda e^{-\lambda t}$$

d) Failure rate:

$$z(t) = \frac{f(t)}{R(t)} = \lambda$$

e) Mean time to failure:

$$MTTF = \int_0^\infty e^{-\lambda t}\, dt = \frac{1}{\lambda}$$

f) System reliability:

$$R_S(t) = e^{-k} \quad (\text{ where } k = t\sum_{i=1}^n \lambda_i)$$

g) Cumulative failure rate:

$$z_S = \lambda_1 + \lambda_2 + \ldots + \lambda_n = \sum_{i=1}^n \lambda_i = \frac{1}{MTBF}$$

PART **III**

PHYSICS

Physics is the science of nature in the broadest sense. Physicists study the behavior and properties of matter in a wide variety of contexts, ranging from the sub-microscopic particles from which all ordinary matter is made (particle physics) to the behavior of the material universe as a whole (cosmology).

This part of the book contains the most frequently-used formulas and definitions related to the following:

1. Mechanics
2. Mechanics of Fluid
3. Temperature and Heat
4. Electricity and Magnetism
5. Light
6. Wave Motion and Sound

MECHANICS

In physics, classical mechanics is one of the two major sub-fields of study in the science of mechanics, (quantum mechanics is the other). Classical mechanics is concerned with the motions of bodies and the forces that cause those motions. This subject concerns macroscopic bodies, i.e., bodies that can be easily seen in the solid state.

This section contains the most frequently used formulas, rules, and definitions related to the following:

1. Kinematics
2. Dynamics
3. Statics

1. Scalars and Vectors

The mathematical quantities that are used to describe the motion of objects can be divided into two categories: scalars and vectors.

a) Scalars

Scalars are quantities that can be fully described by a magnitude alone.

b) Vectors

Vectors are quantities that can be fully described by both a magnitude and a direction.

2. Distance and Displacement

a) Distance

Distance is a scalar quantity that refers to how far an object has gone during its motion.

b) Displacement

Displacement is the change in position of the object. It is a vector that includes the magnitude as a distance, such as five miles, and a direction, such as north.

3. Acceleration

Acceleration is the change in velocity per unit of time. Acceleration is a vector quality.

4.   Speed and Velocity

a) Speed
The distance traveled per unit of time is called the speed, for example 35 miles per hour. Speed is a scalar quantity.

b) Velocity
The quantity that combines both the speed of an object and its direction of motion is called velocity.
Velocity is a vector quantity.

5.   Frequency
Frequency is the number of complete vibrations per unit time in simple harmonic or sinusoidal motion.

6.   Period
Period is the time required for one full cycle. It is the reciprocal of the frequency.

7.   Angular Displacement
Angular displacement is the rotational angle through which any point on a rotating body moves.

8.   Angular Velocity
Angular velocity is the ratio of angular displacement to time.

9. Angular Acceleration

Angular acceleration is the ratio of angular velocity with respect to time.

10. Rotational Speed

Rotational speed is the number of revolutions (a revolution is one complete rotation of a body) per unit of time.

11. Uniform Linear Motion

A path is a straight line. The total distance traveled corresponds with the rectangular area in the diagram $v - t$.

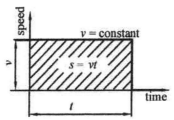

a) Distance:

$$s = vt$$

b) Speed:

$$v = \frac{s}{t}$$

where

> $s =$ distance (m)
> $v =$ speed (m/s)
> $t =$ time (s)

12. Uniform Accelerated Linear Motion

1) If $v_0 > 0; a > 0$, then

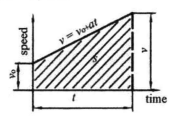

a) Distance:

$$s = v_0 t + \frac{at^2}{2}$$

b) Speed:

$$v = v_0 + at$$

where

> $s =$ distance (m)
> $v =$ speed (m/s)
> $t =$ time (s)
> $v_0 =$ initial speed (m/s)
> $a =$ acceleration (m/s^2)

2) If $v_0 = 0; a > 0$, then

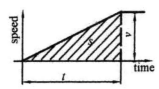

a) Distance:

$$s = \frac{at^2}{2}$$

The shaded areas in diagram $v - t$ represent the distance s traveled during the time period t.

b) Speed:

$$v = a \cdot t$$

where

   s = distance (m)

   v = speed (m/s)

   v_0 = initial speed (m/s)

   a = acceleration (m/s^2)

13. Rotational Motion

Rotational motion occurs when the body itself is spinning. The path is a circle about the axis.

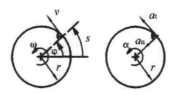

a) Distance:

$$s = r\varphi$$

b) Velocity:

$$v = r\omega$$

c) Tangential acceleration:

$$a_t = r \cdot \alpha$$

d) Centripetal acceleration:

$$a_n = \omega^2 r = \frac{v^2}{r}$$

where

$\widehat{\varphi}$ = angle determined by s and r (rad)

ω = angular velocity $\left(s^{-1}\right)$

α = angular acceleration $\left(1/s^2\right)$

a_t = tangential acceleration $\left(1/s^2\right)$

a_n = centripetal acceleration $\left(1/s^2\right)$

Distance s, velocity v, and tangential acceleration a_t are proportional to radius r.

14. Uniform Rotation about a Fixed Axis

ω_0 = constant; $\alpha = 0$,

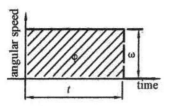

a) Angle of rotation:

$$\varphi = \omega \cdot t$$

b) Angular velocity:

$$\omega = \frac{\varphi}{t}$$

where

φ = angle of rotation (rad)

ω = angular velocity $\left(s^{-1}\right)$

α = angular acceleration $\left(1/s^2\right)$

ω_0 = initial angular speed $\left(s^{-1}\right)$

The shaded area in the diagram $\omega - t$ represents the angle of rotation $\varphi = 2\pi n$ covered during time period t.

15. Uniform Accelerated Rotation about a Fixed Axis

1) If $\omega_0 > 0$; $\alpha > 0$, then

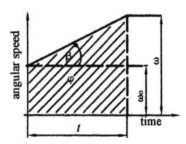

a)   Angle of rotation:

$$\varphi = \frac{1}{2}\left(\omega_0 + \omega\right) = \omega_0 t + \frac{1}{2}\alpha t^2$$

b) Angular velocity:

$$\omega = \omega_0 + \alpha t = \sqrt{\omega_0^2 + 2\alpha\varphi}$$

$$\omega_0 = \omega - \alpha t = \sqrt{\omega^2 - 2\alpha\varphi}$$

c) Angular acceleration:

$$\alpha = \frac{\omega - \omega_0}{t} = \frac{\omega^2 - \omega_0^2}{2\varphi}$$

d) Time: $t = \dfrac{\omega - \omega_0}{\alpha} = \dfrac{2\varphi}{\omega_0 + \omega}$

2) If $\omega_0 = 0;$ $a =$ constant, then

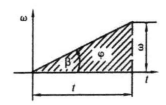

a) Angle of rotation:

$$\varphi = \frac{\omega \cdot t}{2} = \frac{a \cdot t}{2} = \frac{\omega^2}{2a}$$

b) Angular velocity:

$$\omega = \sqrt{2a\varphi} = \frac{2\varphi}{t} = a \cdot t; \; \omega_0 = 0$$

c) Angular acceleration:

$$a = \frac{\omega}{t} = \frac{2\varphi}{t^2} = \frac{\omega^2}{2\varphi}$$

d) Time:

$$t = \sqrt{\frac{2\varphi}{a}} = \frac{\omega}{a} = \frac{2\varphi}{\omega}$$

16. Simple Harmonic Motion

Simple harmonic motion occurs when an object moves repeatedly over the same path in equal time intervals.

The maximum deflection from the position of rest is
called "amplitude."

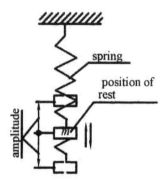

A mass on a spring is an example of an object in simple
harmonic motion.
The motion is sinusoidal in time and demonstrates a
single frequency.

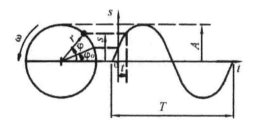

a) Displacement:

$$s = A\sin(\omega \cdot t + \varphi_0)$$

b) Velocity:

$$v = A\omega \cos\left(\omega \cdot t + \varphi_0\right)$$

c) Angular acceleration:

$$a = -A\alpha\omega^2 \sin\left(\omega \cdot t + \varphi_0\right)$$

where

s = displacement

A = amplitude

φ_0 = angular position at time $t = 0$

φ = angular position at time t

T = period

17. Pendulum

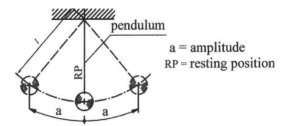

pendulum

a = amplitude
RP = resting position

A pendulum consists of an object suspended so that it swings freely back and forth about a pivot.

a) Period:

$$T = 2\pi \sqrt{\frac{l}{g}}$$

where

T = period (s)
l = length of pendulum (m)
g = 9.81 (m/s^2) or 32.2 (ft/s^2)

18. Free Fall
A free-falling object is an object that is falling due to the sole influence of gravity.

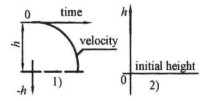

a) Initial speed:

$$v_0 = 0$$

b) Distance:

$$h = -\frac{gt^2}{2} = -\frac{vt}{2} = -\frac{v^2}{2g}$$

c) Speed:

$$v = +gt = -\frac{2h}{t} = \sqrt{-2gh}$$

d) Time:

$$t = +\frac{v}{g} = -\frac{2h}{v} = \sqrt{-\frac{2h}{g}}$$

19. Vertical Projection

a) Initial speed:

$$v_0 > 0, \text{ (upwards)}; \quad v_0 < 0, \text{ (downwards)}$$

b) Distance:

$$h = v_0 t - \frac{gt^2}{2} = (v_0 + v)\frac{t}{2}; \quad h_{max} = \frac{v_0^2}{2g}$$

c) Time:

$$t = \frac{v_0 - v}{g} = \frac{2h}{v_0 + v}; \quad t_{h\,max} = \frac{v_0}{g}$$

where

v = velocity (m/s)

h = distance (m)

g = acceleration due to gravity (m/s^2)

20. Angled Projection

Upwards $(\alpha > 0)$; downwards $(\alpha < 0)$.

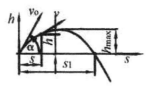

a) Distance:

$$s = v_0 \cdot t \cos\alpha$$

b) Altitude:

$$h = v_0 t \sin\alpha - \frac{g \cdot t^2}{2} = s \tan\alpha - \frac{g \cdot s^2}{2v_0^2 \cos\alpha}$$

$$h_{max} = \frac{v_0^2 \sin^2\alpha}{2g}$$

c) Velocity:

$$v = \sqrt{v_0^2 - 2gh} = \sqrt{v_0^2 + g^2 t^2 - 2gv_0 t \sin\alpha}$$

d) Time:

$$t_{h\max} = \frac{v_0 \sin\alpha}{g}; \quad t_{s1} = \frac{2v_0 \sin\alpha}{g}$$

21. Horizontal Projection $(\alpha = 0)$

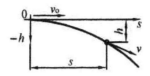

a) Distance:

$$s = v_0 t = v_0 \sqrt{\frac{2h}{g}}$$

b) Altitude:

$$h = -\frac{gt^2}{2}$$

c) Trajectory velocity:

$$v = \sqrt{v_0^2 + g^2 t^2}$$

where

v_0 = initial velocity (m/s)

v = trajectory velocity (m/s)

s = distance (m)

h = height (m)

22. Sliding Motion on an Inclined Plane

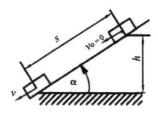

1) If excluding friction ($\mu = 0$), then

   a) Velocity:

$$v = at = \frac{2s}{t} = \sqrt{2as}$$

   b) Distance:

$$s = \frac{at^2}{2} = \frac{vt}{2} = \frac{v^2}{2a}$$

   c) Acceleration:

$$a = g\sin\alpha$$

2) If including friction ($\mu > 0$), then

   a) Velocity:

$$v = at = \frac{2s}{t} = \sqrt{2as}$$

b) Distance:

$$S = \frac{at^2}{2} = \frac{vt}{2} = \frac{v^2}{2a}$$

c) Acceleration:

$$S = \frac{at^2}{2} = \frac{vt}{2} = \frac{v^2}{2a}$$

where

μ = coefficient of sliding friction

g = acceleration due to gravity,

g = 9.81 (m/s^2)

V_0 = initial velocity (m/s)

v = trajectory velocity (m/s)

s = distance (m)

a = acceleration (m/s^2)

α = inclined angle

23. Rolling Motion on an Inclined Plane

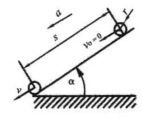

1) If excluding friction $(f = 0)$, then

a) Velocity:

$$v = at = \frac{2s}{t} = \sqrt{2as}$$

b) Acceleration:

$$a = \frac{gr^2}{r^2 + k^2} \sin \alpha$$

c) Distance:

$$s = \frac{at^2}{2} = \frac{vt}{2} = \frac{v^2}{2a}$$

d) Tilting angle:

$$\tan \alpha = \mu_0 \frac{r^2 + k^2}{k^2}$$

2) If including friction $(f > 0)$, then

a) Distance:

$$s = \frac{at^2}{2} = \frac{vt}{2} = \frac{v^2}{2a}$$

b) Velocity:

$$v = at = \frac{2s}{t} = \sqrt{2as}$$

c) Acceleration:

$$a = gr^2 \frac{\sin \alpha - (f/r)\cos \alpha}{r^2 + k^2}$$

d) Tilting angle:

$$\tan \alpha_{min} = \frac{f}{r}; \quad \tan \alpha_{max} = \mu_0 \frac{r^2 + k^2 - fr}{k^2}$$

The value of k can be the calculated by formulas which are given in Table 1.

Table 1 Formulas by calculated radius of gyration (k)

Ball	Solid cylinder	Pipe with low wall thickness
$k^2 = \dfrac{2r^2}{5}$	$k^2 = \dfrac{r^2}{2}$	$k^2 = \dfrac{r_i^2 + r_o^2}{2} \approx r^2$

where

s = distance (m)

v = velocity (m/s)

a = acceleration (m/s^2)

α = tilting angle $\left(^0\right)$

f = lever arm of rolling resistance (m)

k = radius of gyration (m)

μ_0 = coefficient of static friction

g = acceleration due to gravity (m/s^2)

24. Newton's First Law of Motion

Newton's First Law or the Law of Inertia:

An object that is in motion continues in motion with the same velocity at constant speed and in a straight line, and an object at rest continues at rest unless an unbalanced (outside) force acts upon it.

25. Newton's Second Law

The second law of motion, called the Law of Acceleration:

The total force acting on an object equals the mass of the object times its acceleration.

In equation form, this law is

$$F = ma$$

where

F = total force (N)

m = mass (kg)

a = acceleration (m/s^2)

26. Newton's Third Law

The Third Law of Motion, called the Law of Action and Reaction, can be stated as follows:

For every force applied by object A to object B (action), there is a force exerted by object B on object A (the reaction) which has the same magnitude but is opposite in direction.

In equation form this law is

$$F_B = -F_A$$

Dynamics

---

where

F_B = force of action (N)

F_A = force of reaction (N)

27. Momentum of Force

The momentum can be defined as mass in motion. Momentum is a vector quantity; in other words, the direction is important:

$$p = mv$$

28. Impulse of Force

The impulse of a force is equal to the change in momentum that the force causes in an object:

$$I = Ft$$

where

p = momentum (N s)

m = mass of object (kg)

v = velocity of object (m/s)

I = impulse of force (N s)

F = force (N)

t = time (s)

29. Law of Conservation of Momentum

One of the most powerful laws in physics is the law of momentum conservation, which can be stated as follows: *In the absence of external forces, the total momentum of the system is constant.* For example,

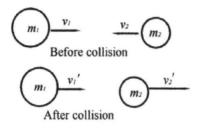

Before collision

After collision

If two objects of mass m_1 and mass m_2, having velocity V_1 and V_2, collide and then separate with velocity V_1' and V_2', the equation for the conservation of momentum is

$$m_1 V_1 + m_2 V_2 = m_1 V_1' + m_2 V_2'$$

30. Friction

Friction is a force that always acts parallel to the surface in contact and opposite to the direction of motion. Starting friction is greater than moving friction. Friction increases as the force between the surfaces increases.

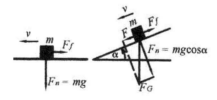

The characteristics of friction can be described by the following equation:

$$F_f = \mu F_n$$

where
   F_f = frictional force (N)
   F_n = normal force (N)
   μ = coefficient of friction ($\mu = \tan\alpha$)

31. General Law of Gravity

Gravity is a force that attracts bodies of matter toward each other. Simply put, gravity is the attraction between any two objects that have mass.

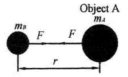

The general formula for gravity is

$$F = \Gamma \frac{m_A m_B}{r^2}$$

where
   m_A , m_B = mass of objects A and B (kg)
      F = magnitude of attractive force
         between objects A and B (N
      r = distance between object A and B(m)
      Γ = gravitational constant (N m^2 / kg^2)
      $\Gamma = 6.67 \times 10^{-11}$ N m^2 / kg^2

32. Gravitational Force

The force of gravity is given by the equation

$$F_G = g \frac{R_e^2 m}{(R_e + h)^2}$$

On the earth surface, $h = 0$; so,

$$F_G = mg$$

where

F_G = force of gravity (N)

R_e = radius of the Earth ($R_e = 6.37 \times 10^6$ m)

m = mass (kg)

g = acceleration due to gravity (m/s^2)

$g = 9.81 \, (m/s^2)$ or $g = 32.2 \, (ft/s^2)$

The acceleration of a falling body is independent of the mass of the object.

The weight F_w on an object is actually the force of gravity on that object:

$$F_W = mg$$

33. Centrifugal Force

Centrifugal force is the apparent force drawing a rotating body away from the center of rotation, and it is caused

Dynamics

by the inertia of the body. Centrifugal force can be calculated by the formula:

$$F_c = \frac{mv^2}{r} = m\omega^2 r$$

34. Centripetal Force

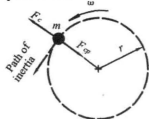

Centripetal force is defined as the force acting on a body in curvilinear motion that is directed toward the center of curvature or axis of rotation. Centripetal force is equal in magnitude to centrifugal force but in the opposite direction.

$$F_{cp} = -F_c = -\frac{mv^2}{r}$$

where
F_c = centrifugal force (N)
F_{cp} = centripetal force (N)
m = mass of the body (kg)
v = velocity of the body (m/s)
r = radius of curvature of the path of the body (m)
ω = angular velocity $\left(s^{-1}\right)$

35. Torque

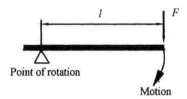

Torque is the ability of a force to cause a body to rotate about a particular axis.

Torque can have either a clockwise or a counterclockwise direction. To distinguish between the two possible directions of rotation, we adopt the convention that a counterclockwise torque is positive and that a clockwise torque is negative.

One way to quantify a torque is

$$T = F \cdot l$$

where
T = torque (N m or lb ft)
F = applied force (N or lb)
l = length of torque arm (m or ft)

36. Work

Work is the product of a force in the direction of the motion and the displacement.

Dynamics

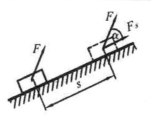

a) Work done by a constant force:

$$W = F_s \cdot s = F \cdot s \cdot \cos \alpha$$

where
   W = work (Nm = J)
   F_s = component of force along the direction of
      movement (N)
   s = distance the system is displaced (m)

b) Work done by a variable force
If the force is not constant along the path of the object,
we need to calculate the force over very tiny intervals
and then add them up. This is exactly what the
integration over differential small intervals of a line can
accomplish:

$$W = \int_{si}^{sf} F_s(s) \cdot ds = \int_{si}^{sf} F(s) \cos \alpha \cdot ds$$

where
   $Fs(s)$ = component of the force function along the
      direction of movement (N)

$F(s)$ = function of the magnitude of the force vector along the displacement curve (N)

S_i = initial location of the body (m)

S_f = final location of the body (m)

α = angle between the displacement and the force

37. Energy

Energy is defined as the ability to do work. The quantitative relationship between work and mechanical energy is expressed by the equation:

$$TME_i + W_{ext} = TME_f$$

where

TME_i = initial amount of total mechanical energy (J)

W_{ext} = work done by external forces (J)

TME_f = final amount of total mechanical energy (J)

There are two kinds of mechanical energy: kinetic and potential.

a) Kinetic energy

Kinetic energy is the energy of motion. The following equation is used to represent the kinetic energy of an object:

$$E_k = \frac{1}{2}mv^2$$

where

m = mass of moving object (kg)

v = velocity of moving object (m/s)

b) Potential energy

Potential energy is the stored energy of a body and is due to its internal characteristics or its position. Gravitational potential energy is defined by the formula

$$E_{pg} = m \cdot g \cdot h$$

where

E_{pg} = gravitational potential energy (J)

m = mass of object (kg)

h = height above reference level (m)

g = acceleration due to gravity (m/s^2)

38. Conservation of Energy

In any isolated system, energy can be transformed from one kind to another, but the total amount of energy is constant (conserved):

$$E = E_k + E_p + E_e + \ldots = \text{constant}$$

Conservation of mechanical energy is given by

$$E_k + E_p = \text{constant}$$

39. Relativistic Energy

It is a consequence of relativity that the energy of a particle of rest mass m moving with speed v is given by

$$E = \frac{mc^2}{\sqrt{1 - \dfrac{v^2}{c^2}}}$$

where

m = rest mass of the body
v = velocity of the body (m/s).
c = speed of light, $c = 3 \times 10^8$ m/s

$$\frac{1}{\sqrt{1 - \dfrac{v^2}{c^2}}} = \text{Lorentz factor}$$

40. Power

Power is the rate at which work is done, or the rate at which energy is transformed from one form to another. Mathematically, it is computed using the following equation:

$$P = \frac{W}{t}$$

where

P = power (W)
W = work (J)
t = time (s)

The standard metric unit of power is the watt (W). As is implied by the equation for power, a unit of power is equivalent to a unit of work divided by a unit of time. Thus, a watt is equivalent to Joule/second (J/s). Since the expression for work is

$$W = F \cdot s,$$

the expression for power can be rewritten as

$$P = F \cdot v$$

where

s = displacement (m)
v = speed (m/s)

41. Resolution of a Force

$$F_x = F\cos\alpha; \quad F_y = F\sin\alpha$$

$$F = \sqrt{F_x^2 + F_y^2}; \quad \tan\alpha = \frac{F_y}{F_x}$$

42. Moment of a Force about a Point 0

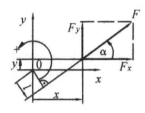

$$M_0 = \pm Fl = F_y x - F_x y.$$

43. Mechanical Advantage of Simple Machines

The mechanical advantage is the ratio of the force of resistance to the force of effort:

$$MA = \frac{F_R}{F_E}$$

where

$\quad MA$ = mechanical advantage

$\quad F_R$ = force of resistance (N)

$\quad F_E$ = force of effort (N)

44. The Lever

A lever consists of a rigid bar that is free to turn on a pivot, which is called a fulcrum.

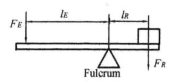

The law of simple machines as applied to levers is

$$F_R \cdot l_R = F_E \cdot l_E$$

45. Wheel and Axle

A wheel and axle consist of a large wheel attached to an axle so that both turn together:

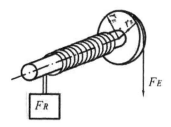

$$F_R \cdot r_R = F_E \cdot r_E$$

where

F_R = force of resistance (N)

F_E = force of effort (N)

r_R = radius of resistance wheel (m)

r_E = radius of effort wheel (m)

The mechanical advantage is

$$MA_{\text{wheel and axle}} = \frac{r_E}{r_R}$$

46. The Pulley

If a pulley is fastened to a fixed object, it is called a *fixed pulley*. If the pulley is fastened to the resistance to be moved, it is called a *movable pulley*. When one continuous cord is used, the ratio reduces according to the number of strands holding the resistance in the pulley system.

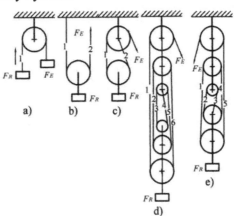

The effort force equals the tension in each supporting strand. The mechanical advantage of the pulley is given by formula:

$$MA_{\text{pulley}} = \frac{F_R}{F_E} = \frac{nT}{T} = n$$

where

T = tension in each supporting strand

N = number of strands holding the resistance

F_R = force of resistance (N)

F_E = force of effort (N)

47. The Inclined Plane

An inclined plane is a surface set at an angle from the horizontal and used to raise objects that are too heavy to lift vertically:

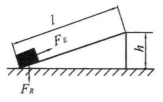

The mechanical advantage of an inclined plane is

$$MA_{\text{inclined plane}} = \frac{F_R}{F_E} = \frac{l}{h}$$

where

F_R = force of resistance (N)

F_E = force of effort (N)

l = length of plane (m)

h = height of plane (m)

48. The Wedge

The wedge is a modification of the inclined plane.
The mechanical advantage of a wedge can be found by dividing the length of either slope by the thickness of the longer end.

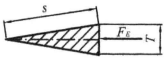

As with the inclined plane, the mechanical advantage gained by using a wedge requires a corresponding increase in distance.

The mechanical advantage is:

$$MA = \frac{s}{T}$$

where:

MA = mechanical advantage

s = length of either slope (m)

T = thickness of the longer end (m)

Statics

---

49. The Screw

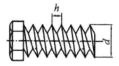

A screw is an inclined plane wrapped around a circle.
From the law of machines,

$$F_R \cdot h = F_E \cdot U_E$$

However, for advancing a screw with a screwdriver, the
mechanical advantage is:

$$MA_{\text{screw}} = \frac{F_R}{F_E} = \frac{U_E}{h}$$

where

F_R = force of resistance (N)

F_E = effort force (N)

h = pitch of screw

U_E = circumference of the handle of the screw

MECHANICS OF FLUIDS

The branch of mechanics called "mechanics of fluids" is concerned with fluids, which may be either liquids or gases. This topic involves various properties of fluids, such as velocity, pressure, density and temperature, as functions of space and time. Typically, liquids are considered to be uncompressible, whereas gases are considered to be compressible.

This section of the book contains the most frequently used formulas and definitions relating to hydrostatics and hydrodynamics.

1. Density

Density is the ratio of mass to volume:

$$\rho = \frac{m}{V}$$

where

ρ = density (kg/m^3)

m = mass (kg)

V = volume (m^3)

2. Viscosity

Viscosity is the measure of the internal friction between the molecules of liquid that resist motion across each other.

a) Dynamic viscosity

The dynamic viscosity is a material constant which is a function of pressure and temperature:

$$\eta = f(p, t)$$

b) Kinematic viscosity:

$$v = \frac{\eta}{\rho}$$

where

v = kinematic viscosity $\left(m^2/s\right)$

ρ = density (kg/m^3)

η = dynamic viscosity (Pa s)

$$1\,Pa\ s = \frac{kg}{m\ s} = \frac{N\ s}{m^2} = 10P$$

Viscosity measurements are expressed in "Pascal-seconds" (Pa s) or "milli-Pascal-seconds" (mPa s); these are units of the International System and are sometimes used in preference to the metric designations. But the most frequently used unit of viscosity measurement is the "poise" (P). (A material requiring a shear stress of one dyne per square centimeter to produce a shear rate of one reciprocal second has a viscosity of one poise, or 100 centipoise).

One Pascal-second is equal to ten poise; one milli-Pascal-second is equal to one centipoise.

3. Pressure of Solid

Pressure is force applied to a unit area:

$$p = \frac{F}{A}$$

where

p = pressure (N/m^2 or lb/in^2)

F = force applied (N or lb)

A = area, $(m^2 \text{ or } in^2)$.

$1\,N/m^2 = 1\,Pa$

4. Pressure of Liquids

Pressure in liquid depends only on the depth and density of the liquid and not on the surface area.

The pressure at any depth is, however, due not only to the weight of liquid above but to the pressure of air above the surface as well. The total pressure at a depth h is therefore given by the sum of these two pressures.

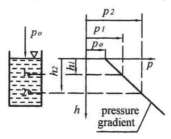

a) Pressure at a depth h_0

Pressure at the free surface of the liquid $(h = 0)$ is only the air pressure:

$$p_0 = p_a$$

b) Pressure at a depth h_1:

$$p_1 = p_0 + g\rho h_1$$

c) Pressure at a depth h_2:

$$p_2 = p_1 + g\rho(h_2 - h_1) = p_0 + g\rho h_2$$

where

p_1, p_2 = pressure on a depth 1 and 2 (Pa)

h_1, h_2 = depth 1 and 2 (m)

p_a = air pressure (Pa)

p_0 = pressure on a free surface of
the liquid (Pa)

ρ = density of the liquid (kg/m^3)

g = acceleration due to gravity (m/s^2)

5. Force Exerted by Liquids

a) Force on a horizontal surface

The force exerted by a liquid on a horizontal surface is given by the formula

$$F = g\rho h A_h$$

where

A_h = area of horizontal surface (m^2)

h = depth of the liquid (m)

ρ = density of the liquid (kg/m^3)

g = acceleration due to gravity (m/s^2)

b) Force on a vertical surface:

The force on a vertical surface is found by using half the vertical height and is given by the formula

$$F_s = \frac{1}{2} g\rho h_a A_s$$

where

A_s = area of the side or vertical surface (m^2)

h_a = average depth of the liquid (m)

ρ = density of the liquid (kg/m^3)

g = acceleration due to gravity (m/s^2)

6. Pascal's Principle

Pressure exerted on an enclosed liquid is transmitted equally to every part of the liquid and to the walls of the container. Pascal's principle is important in understanding hydraulics, which is the study of the transfer of forces through fluids.

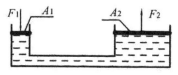

$$p = \frac{F_1}{A_1} = \frac{F_2}{A_2}$$

where

A_1, A_2 = area of small and large cylinders (m^2)

F_1, F_2 = applied and upward forces (N)

---

7. Archimedes' Principle

Any object placed in a fluid apparently loses weight equal to the weight of the displaced fluid.

For water, which has a density $\rho_w = 1 \text{g/cm}^3$, this provides a convenient way to determine the volume of an irregularly shaped object and then to determine its density:

$$m_o - m_{app} = \rho_w \cdot V_o$$

where

m_o = mass of object (kg)

m_{app} = apparent mass when submerged (kg)

V_o = volume of object (m^3)

ρ_w = density of the water (kg/m^3)

8. Buoyant Force

When a rigid object is submerged in a fluid, there exists a buoyant force (an upward force) on the object that is equal to the weight of the fluid that is displaced by the object. This force is given by the equation:

$$F_b = \rho g V$$

where

F_b = buoyant force (N)

ρ = density of the liquid (kg/m^3)

g = acceleration due to gravity (m/s^2)

V = volume of submerged object (m^3)

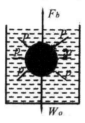

The net force on the object is given by

$$F_n = F_b - W_o = \rho_f \cdot V_s \cdot g - \rho_o \cdot V_o \cdot g$$

where

F_n = net force on object (N)

F_b = buoyant force (N)

W_o = weight of the object (kg)

ρ_f = density of the fluid (kg/m^3)

V_s = volume of submerged (m^3)

ρ_o = density of the object (kg/m^3)

V_o = volume of the object (m^3)

g = acceleration due to gravity (m/s^2)

When an object is floating, the net force on it will be zero. This happens when the volume of the object submerged displaces an amount of liquid whose weight is equal to the weight of the object. A ship made of steel can float because it can displace more water than it weighs.

9. Flow Rate

The flow rate of a fluid is the volume of fluid flowing past a given point in a pipe per unit time:

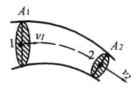

$$Q = A_1 \cdot v_1 = A_2 \cdot v_2 = \text{constant}$$

where

Q = flow rate (m^3 / s)

v_1, v_2 = flow velocity at point 1 and point 2 (m/s)

A_1, A_2 = cross-sectional area at sections 1 and 2 (m^2)

10. Conservation of Mass

The rate of mass that goes into a system is equal to the rate of accumulation plus the rate of mass that goes out.

At steady (lamellar) state, the rate of accumulation is zero; therefore

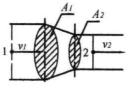

$$A_1 v_1 \rho_1 = A_2 v_2 \rho_2 = A v \rho$$

where

A_1, A_2 = areas of the pipe-cross section at point 1 and point 2 (m^2)

V_1 = fluid velocity at point 1 (m/s)

V_2 = fluid velocity at point 2 (m/s)

ρ_1 = density of fluid at point 1 (kg/m^3)

ρ_2 = density of fluid at point 2 (kg/m^3)

11. Bernoulli's Equation

Bernoulli's equation is based on the concept that points 1 and 2 lie on a streamline, the fluid has constant density, the flow is steady, and there is no friction.

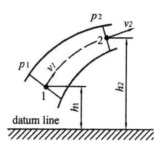

$$p_1 + h_1 \rho g + \frac{1}{2} \rho v_1^2 = p_2 + h_2 \rho g + \frac{1}{2} \rho v_2^2$$

where

p_1 = fluid pressure at point 1 (Pa)

p_2 = fluid pressure at point 2 (Pa)

v_1 = fluid velocity at point 1 (m/s)

v_2 = fluid velocity at point 2 (m/s)

h_1 = elevation at point 1 (m)

h_2 = elevation at point 2 (m)

g = acceleration due to gravity (m/s^2)

TEMPERATURE AND HEAT

Thermodynamics is a branch of physics. It is the study of the effects of work, heat, and energy on systems. Heat is a form of energy transferred from one body or system to another as a result of a difference in temperature. The energy associated with the motion of atoms or molecules is capable of being transmitted through solid and fluid media by conduction, through fluid media by convection, and through empty space by radiation. Temperature is the specific degree of hotness or coldness of a body or an environment. It is usually measured with a thermometer or other instrument having a scale calibrated in units (degrees).

This section contains the most frequently used formulas, rules, and definitions relating to

1. Thermal Variables of State
2. Temperature and Heat
3. Changes of State
4. Gas Laws
5. Laws of Thermodynamics.

Thermal Variables of State

---

1. Pressure

The pressure of a system is defined as the force exerted by the system on the unit area of its boundaries. This is the definition of absolute pressure. A state of pressure means $p_g > 0$, and a vacuum means $p_g < 0$. Thus, absolute pressure can be expressed by

$$p = p_g + p_0$$

where

p = absolute pressure (Pa)

p_g = gauge pressure (Pa)

p_0 = atmospheric pressure (Pa)

2. Temperature

Basically, temperature is a measure of the hotness or coldness of the object. There are four basic temperature scales: Celsius $\left(^0C \right)$, Kelvin (K), Fahrenheit (^0F), and Rankine (^0R).

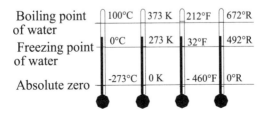

The Kelvin scale is closely related to the Celsius scale

$$T_K = t_C + 273^0$$

The Rankine scale is closely related to the Fahrenheit scale

$$t_R = t_F + 460^0$$

The relationship between Celsius temperatures and Fahrenheit temperatures is given by

$$t_C = \frac{5}{9}\left(t_F - 32^0\right); \quad t_F = \frac{9}{5}t_C + 32^0$$

3. Density

Density is measurement of mass per unit of volume:

$$\rho = \frac{m}{V}$$

where

ρ = object's density (kg/m^3)

m = object's total mass (kg)

V = object's total volume (m^3)

4. Specific Volume

Specific volume is the volume per unit mass or the inverse of density:

$$v = \frac{V}{m} = \frac{1}{\rho}$$

where

v = specific volume (m^3/kg)

ρ = object's density (kg/m^3)

m = object's total mass (kg)

V = object's total volume (m^3)

5. Molar Mass

Molar mass is the mass of one mole of a substance.

a) Mass of one molecule:

$$m_M = M_r \cdot u$$

where

u = unified atomic mass ($u = 1.66 \times 10^{-27}$ kg)

M_r = relative molecular mass

The relative molecular mass of a substance is equal to the relative atomic mass of its constituent atoms.

b) Molar mass of a substance:

$$M = \frac{m}{n} = M_r \cdot N_o$$

where

m = a mass of the substance (g)

n = number of moles of the substance (mol)

N_A = Avogadro's number (mol^{-1})

6. Molar Volume

The molar volume is the volume occupied by one mole of ideal gas at standard temperature and pressure (STP).

a) Standard temperature:
$$T_0 = 273.15 K = 0^0 C$$

b) Standard pressure:
$$p_0 = 101325 \, \text{Pa} = 1.03 \, \text{bar}$$

c) Molar volume value:
$$V_m = 2.24 \times 10^{-2} \, \text{m}^3 \text{mol}^{-1}$$

d) Volume of a gas
$$V = nV_m$$

7. Heat

Heat is the energy that flows spontaneously from a higher temperature object to a lower temperature object through random interactions between their atoms. Heat energy exchanged between objects of different temperatures is expressed as
$$Q = mc(T_2 - T_1)$$

where

Q = heat thermal energy (J)

m = object's total mass (kg)

c = specific heat (J/kg K)

T_2 = temperature of the hot object (K)

T_1 = temperature of the cool object (K)

---

8. Specific Heat

The specific heat is the amount of heat per unit mass required to raise the temperature by one degree Celsius:

$$c = \frac{Q}{m\Delta T}$$

9. Heat Conduction

The total amount of heat transfer between two plane surfaces is given by the equation

$$Q = \frac{kAt(T_2 - T_1)}{l}$$

where

Q = heat transferred (J or Btu)

k = thermal conductivity $\left(\text{J/s m }^0\text{C}\right)$

A = plane area (m^2)

l = thickness of barrier (m)

T_2 = temperature of the hot side (K)

T_1 = temperature of the cool side (K)

10. Expansion of Solid Bodies

a) Linear expansion:

The amount that a solid expands can be written by formula

$$\Delta l = \alpha l \Delta T$$

where

Δl = change in length (m)

l = original length (m)

α = coefficient of linear expansion $\left(\text{m} / \, ^0\text{C} \right)$

ΔT = change in temperature

b) Area and volume expansion:
To allow for this expansion, the following formulas are used:

$$\Delta A = 2\alpha A \Delta T$$

$$\Delta V = 3\alpha V \Delta T$$

where

A = original area $\left(\text{m}^2 \right)$

V = original volume $\left(\text{m}^3 \right)$

11. Expansion of Liquids

The formula for volume expansion of liquids is

$$\Delta V = \beta V \Delta T$$

where

V = original volume $\left(\text{m}^3 \right)$

β = coefficient of volume expansion for liquids

12. Expansion of Water

The most common liquid, water, does not behave like other liquids. Above about 4°C, water expands as the temperature rises, as we would expect. Between 0 and about 4°C, however, water *contracts* with increasing temperature. Thus, at exactly 3.98°C, the density of water passes through a maximum. At all other temperatures, the density of water is less than this maximum value.

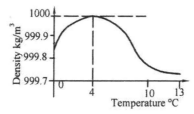

13. Fusion

The change of state from solid to liquid is called fusion or melting. The change from the liquid to the solid is called freezing or solidification. The heat of fusion L_f is the quantity of heat energy required to convert one mass unit of solid to liquid:

$$L_f = \frac{Q}{m}$$

where

Q = quantity of heat (J)
m = mass (kg)

14. Vaporization

The change of state from a liquid to a gaseous or vaporous state is called vaporization.

The heat of vaporization Lv is the heat required to vaporize one mass unit of a substance at its normal boiling point:

$$L_v = \frac{Q}{m}$$

15. Equation of State

The equation of state of a gas in thermal equilibrium relates the pressure, the volume, and the temperature of a gas. All gases have the same equation of state, called the ideal gas law:

$$pV = NkT = nRT$$

where

N = number of molecules in the gas

n = number of moles of the gas (mol)

T = Kelvin temperature of the gas (K)

p = pressure (Pa)

V = volume $\left(m^3\right)$

k = Boltzmann's constant

   ($k = 1.38 \times 10^{-23}$ J/K)

R = universal gas constant ($R = 8.314$ J/mol $\cdot$ K)

The ratio,

$$N_A = \frac{R}{k} = 6.022 \times 10^{23} \text{ mol}^{-1}$$

is Avogadro's number, which is the number of molecules in a mole.

16. The Charles Law for Temperature

If the pressure on a gas is constant, $p =$ constant, the volume is directly proportional to its absolute temperature:

$$\frac{V_1}{T_1} = \frac{V_2}{T_2} \quad \text{or} \quad V_1 T_2 = V_2 T_1$$

where

V_1 = original volume (m^3)

T_1 = original temperature (K)

V_2 = final volume (m^3)

T_2 = final temperature (K)

17. Boyle's Law for Pressure

If the temperature of the gas is constant, $T =$ constant, the volume is inversely proportional:

$$p_1 V_1 = p_2 V_2$$

where

V_1 = original volume (m^3)

p_1 = original pressure (Pa)

V_2 = final volume (m^3)

p_2 = final pressure (Pa)

18. Gay-Lussac's Law for Temperature

The pressure of a given mass of gas is directly proportional to the Kelvin temperature if the volume is kept constant:

$$\frac{p_1}{T_1} = \frac{p_2}{T_2}$$

where

p_1 = original pressure (Pa)

p_2 = final pressure (Pa).

T_1 = original temperature (K)

T_2 = final temperature (K)

19. Dalton's Law of Partial Pressures

At constant volume and temperature, the total pressure (p_T) exerted by a mixture of gases is equal to the sum of the partial pressures:

$$p_T = p_1 + p_2 + p_3 + ... + p_n$$

where

p_T = total pressure (Pa)

$p_1 + p_2 + p_3 + ... + p_n$ = partial pressures (Pa)

20. Combined Gas Law

Most of the time, it is very difficult to keep pressure or temperature constant. To keep these parameters constant, the best solution is to combine Charles' law and Boyle's as follows:

$$\frac{p_1 V_1}{T_1} = \frac{p_2 V_2}{T_2}$$

21. The First Law of Thermodynamics

The first law of thermodynamics is often called the law of conservation of energy. This law suggests that energy can be transformed from one kind of matter to another in many forms. However, it cannot be created nor destroyed.

The first law of thermodynamics defines internal energy (E) as equal to the heat transfer (Q) *into* a system and the work (W) done by the system.

$$E_2 - E_1 = \Delta E = Q - W$$

where

ΔE = change in internal energy
Q = heat added into the system
W = work done by the system

22. The Second Law of Thermodynamics

In physics, the second law of thermodynamics, in its many forms, is a statement about the quality and direction of energy flow, and it is closely related to the concept of entropy. This law suggests that heat can never pass spontaneously from a colder to a hotter body. As a result of this fact, natural processes that involve energy transfer must have one direction, and all natural processes are irreversible.

a) Entropy:

Thermodynamic entropy (S) is a measure of the amount of energy in a physical system that cannot be used to do work. A state variable whose change is defined for a reversible process at temperature T and the heat absorbed Q. The entropy change is

$$\Delta S = \frac{Q}{T}$$

where

ΔS = entropy change (J/K)
Q = heat absorbed (J)
T = temperature (K)

The importance of the entropy function is exhibited in the second law of thermodynamics.
In any process, the total entropy of the system and its surroundings increases or (in a reversible process) does not change.

b) Heat engines and refrigerators:
A heat engine is a device or system that converts heat into work. The efficiency of a cyclic heat engine is

$$\eta = \frac{W}{Q_h} = 1 - \frac{Q_c}{Q_h}$$

where

Q_h = heat absorbed per cycle from the higher
     temperature (J)

Q_c = heat rejected per cycle to the lower
     temperature (J)

W = work carried out per cycle (J)

The most efficient heat engine cycle is the Carnot cycle, consisting of two isothermal processes and two adiabatic processes.

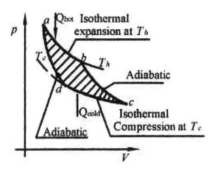

This maximum thermal efficiency is

$$\eta = 1 - \frac{T_c}{T_h}, \quad (T_h > T_c), \text{ but also}$$

$$\eta = 1 - \frac{Q_{\text{cold}}}{Q_{\text{hot}}}$$

23. The Third Law of Thermodynamics

The third law of thermodynamics states that the entropy of a system at zero absolute temperature is a well-defined constant.

Absolute zero = 0 K = -273.15 $^0 C$

ELECTRICITY AND MAGNETISM

Electricity is electrical charge. Franklin, Faraday, Thompson, Einstein, Tesla, and many other historical scientists used the word "electricity" in this way, stating that an electric current is a flow of electricity.

Magnetism is a force that acts at a distance and is caused by a magnetic field. This force strongly attracts ferromagnetic materials such as iron, nickel and cobalt. In magnets, a magnetic force strongly attracts the opposite pole of another magnet and repels the like pole. A magnetic field is both similar and different to an electric field.

This section contains the most frequently used formulas, rules, and definitions regarding to the following:

1. Electrostatics
2. Direct current
3. Magnetism
4. Alternating current

Electrostatics

---

1. Coulomb's Law

The force between two point charges Q_1 and Q_2 is directly proportional to the product of their magnitudes and inversely proportional to the square of the distance separating them, r.

In equation form, Coulomb's law is

$$F = \frac{kQ_1Q_2}{r^2}$$

where

F = force of attraction or repulsion (N)

k = constant, $\left(k = 8.99 \times 10^9 \, \text{Nm}^2/\text{C}^2 \text{for air}\right)$

Q_1, Q_2 = size of charges in coulombs (C)

r = distance between the charges (m)

2. Electric Fields

Electric field strength is a vector quantity having both magnitude and direction.

The magnitude of the electric field of point charge is simply defined as the force per charge of the test charge:

$$E = \frac{F}{q}$$

where

E = electric field strength (N/C)

q = quantity of charge of the test charge (C)
F = electric force (N).

When applied to two charges the source charge Q and the test charge q, the formula for electric force can be written as

$$E = \frac{kQ}{r^2}$$

a) The principle of superposition for electric fields: The total electric field at any point, made up of a distribution of charges $q_1, q_2, q_3, \cdots, q_n$, is found by adding the fields independently established at that point by the individual charges:

$$E_{total} = E_1 + E_2 + E_3 + \dots + E_n$$

3.   Electric Flux

The electric flux is the product of the components of the electric field that are perpendicular to the surface, times the surface area:

$$\Phi_E = EA\cos\theta$$

where

Φ_E = electric flux $\left(\text{Nm}^2 / \text{C}\right)$

θ = angle between field and area vector

A = area vector $\left(\text{m}^2\right)$

E = electric field (N/C)

4. Gauss' Law

The electric flux through any closed surface is equal to the charge inside that surface, divided by a constant ε_0:

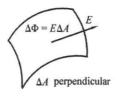

$$\Phi_E = \oint \vec{E}d\vec{A} = \frac{q_{inc}}{\varepsilon_0}$$

where

Φ_E = total electric flux

ε_0 = permittivity of free space constant

$\varepsilon_0 = 8.854 \times 10^{-12} \left(\text{C}^2 / \text{Nm}^2\right)$

q_{inc} = sum of all the enclosed charges

These equations apply in a vacuum and for the most part also in air.

5. Electric Potential

Electric potential can be stated as potential energy per unit charge.

The electric potential V at a distance r from a charge q is

$$V = k\frac{q}{r}$$

where

V = electrical potential (V)
k = constant $\left(k = 8.99 \times 10^9 \, \text{Nm}^2/\text{C}^2\right)$
q = charge (C)
r = distance (m)

a) Principle of superposition of electric potential: When more than one charge is present, the electric potential at a given point is the algebraic sum of the potentials due to each of the charges present. The electric potential V at any point is given by

$$V = V_1 + V_2 + V_3 + ... + V_n = \sum \frac{q_i}{r_i}$$

where

q_i = charge
r_i = distance of the charge
V_n = potential due to n different charges

---

6. Electric Potential Energy

The electric potential can also be defined as the electric potential energy per unit charge. Hence,

$$V = \frac{U}{q} = \frac{W}{q}$$

where

U = magnitude of electric potential energy
W = work done
q = charge

7. Capacitance

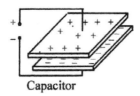

Capacitor

Capacitance is a measure of the amount of stored electric charge for a given electric potential:

$$C = \frac{Q}{V}$$

where

C = capacitance (F)
Q = total electric charge (C)
V = electric potential (V)

8. Capacitor

The capacitance of a capacitor can be calculated by the following formula:

$$C = \varepsilon_0 \varepsilon_r \frac{A}{d}$$

where

C = capacitance (F)

ε_0 = permittivity of free space (F/m)

ε_r = dielectric constant of the insulator (F/m)

A = area of each electrode plate (m^2)

d = distance between the electrodes (m^2)

a) Capacitances in parallel

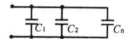

The equivalent capacitance of capacitors connected in parallel is

$$C_{eq} = C_1 + C_2 + ... + C_n$$

b) Capacitances in series

The equivalent capacitance of capacitors connected in series is

$$\frac{1}{C_{eq}} = \frac{1}{C_1} + \frac{1}{C_2} + ... + \frac{1}{C_n}$$

c) Energy

The energy stored in a capacitor is equal to the *work* done to charge it up

$$W_{sto} = \frac{1}{2}CV^2$$

where

W_{sto} = energy stored in capacitor (J)

C = capacitance (F)

V = electric potential (V)

9. Electric Current

The rate of flow of electrons through a conductor from a negatively charged area to one that has a positive charge is called direct current. Thus,

$$I = \frac{Q}{t}$$

where

I = current (A)

Q = charge (C)

t = time (s)

---

10. Current Density

The magnitude of a current's density is the current through a unit area perpendicular to the flow direction. Thus,

$$J = \frac{I}{A}$$

where

J = current density $\left(A/m^2\right)$

A = conductor's cross-section area $\left(m^2\right)$

I = current (A)

11. Potential Difference

The electric potential difference (V) is the work done per unit charge as a charge is moved between two points a and b in an electric field

$$V_a - V_b = V = \frac{W_{ab}}{Q}$$

where

V = electric potential difference (V)

W_{ab} = work as a charge moved between

points a and b (J)

Q = charge (C)

12. Resistance

Resistance is the feature of a material that determines the flow of electric charge:

$$R = \rho \frac{l}{A}$$

where

> R = resistance (Ω)
> l = length (m)
> A = cross-section area (m^2).
> ρ = resistivity, a constant t, which depends on the
>   type of material ($\Omega \cdot$m)

Very often one specifies, instead of ρ, the conductivity

$$\sigma = \frac{1}{\rho}$$

where

> σ = conductivity (S/m)

13. Ohm's Law

The current I in a "resistor" is very nearly proportional
to the difference V in electric potential between the ends
of the resistor. This proportionality is expressed by
Ohm's law:

$$V = IR \quad \text{or} \quad I = \frac{V}{R}$$

where

> I = current through the resistance (A)
> V = potential difference (V)
> R = resistance (Ω)

14. Series Circuits

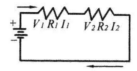

a) Potential difference

The total potential difference is the sum of the potential difference of each component:

$$V = V_1 + V_2 + \cdots + V_n$$

b) Resistance

The total resistance is equal to the sum of the resistance of each component:

$$R = R_1 + R_2 + \cdots + R_n$$

c) Current

The total current is equal in every component.

$$I = I_1 = I_2 = \cdots = I_n$$

15. Parallel Circuits

---

a) Potential difference

The total potential difference is equal in every component.

$$V = V_1 = V_3 = V_3 = \cdots + V_n$$

b) Resistance

The resistance is equal to the sum of resistance of each component divided by the product of the resistance of each component:

$$\frac{1}{R} = \frac{1}{R_1} + \frac{1}{R_2} + \frac{1}{R_3} + \cdots + \frac{1}{R_n}$$

c) Current

The total current is equal to the sum of the current in each component:

$$I = I_1 + I_2 + I_3 + \cdots + I_n$$

16. Series-Parallel Circuit

Many circuits are both series and parallel.

a) Potential difference

The total potential difference is the potential difference of series circuit plus the potential difference of parallel circuits.

$$V = V_1 + V_2 = V_1 + V_3$$

b) Resistance

The total resistance is the resistance of the series circuit plus the resistance of the parallel circuits.

$$R = R_1 + \frac{R_2 R_3}{R_2 + R_3}$$

c) Current

The total current is equal to the current of the series circuit and to the sum of the current of the parallel circuits.

$$I = I_1 = I_2 + I_3$$

17. Joule's Law

a) Work

The "work" or heat energy produced by a resistor is

$$W = I^2 Rt = \frac{V^2}{R} t$$

where

W = work energy or heat energy (J)

I = current (A)
R = resistance (Ω)
V = potential difference (V)
t = time (s)

b) Power

Electrical power is defined as the time rate of doing work. The power consumption of a resistor is

$$P = VI = I^2 R = \frac{V^2}{R}$$

where

P = power (W)
I = current (A)
R = resistance (Ω)
V = potential difference (V).

18. Kirchhoff's Junction Law

For a given junction or node in a circuit, the sum of the currents entering equals the sum of the currents leaving it. In other words, the algebraic sum of all the currents in the junction is zero (as, for example, $I_1 + I_2 = I_3$.) In this case, a current going out of the junction is counted as negative.

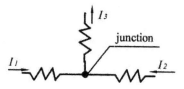

That is, at any junction,

$$\sum_{j=1}^{n} I_j = 0$$

19. Kirchhoff's Loop Law

The algebraic sum of the potential changes around any complete loop in the network is zero, so the sum of the voltage drops equals the voltage source.

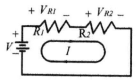

In this example,

$$V = V_{R1} + V_{R2}$$

That is, at any complete loop,

$$\sum_{loop} V = 0$$

---

20. Resistors

Electrical components called *resistors* are used to limit or set current in a circuit with a given voltage, or to set voltage for a given current. (A circuit *element* is an idealization of an actual electronic part, or *component.*) Resistors are usually marked with at least three color bands that indicate their resistance in units of ohms Ω. For 5% tolerance resistors, the first two bands are the first two significant digits of the value, and the third band is the number of zeros to be added to the first two digits. A final band of gold (5%) or silver (10%) indicates the tolerance. For 1% resistors, the first three bands are the first three digits; the fourth is the multiplier. The color code is:
BLACK 0, BROWN 1, RED 2, ORANGE 3, YELLOW 4, GREEN 5, BLUE 6, VIOLET 7, GRAY 8, WHITE 9.

21. Internal Resistance

A cell has resistance within itself, which opposes the movement of electrons. This is called the internal resistance. The voltage applied to the external circuit is, then,

$$V = E - I \cdot r$$

where

V = voltage applied to circuit (V)
E = potential difference across a source (V)

I = current through cell (A)
r = internal resistance of cell (Ω)

22. Magnetic Forces on Moving Charges

A magnetic field is an entity produced by moving electric charges exerting a force on other moving charges. The following equation describes force:

$$F = qvB\sin\theta$$

where

F = force (N)
q = electric charge (C)
v = velocity of the charge (m/s)
B = strength of the magnetic field (T)
θ = smaller angle between the vectors v and B

$$1\,\text{T} = 1\frac{\text{N}\cdot\text{s}}{\text{C}\cdot\text{m}} = 1\frac{\text{N}}{\text{A}\cdot\text{m}} = \frac{\text{V}\cdot\text{s}}{\text{m}^2}$$

23. Force on a Current-Carrying Wire

If instead of a moving charge such as an electron or proton, there is electric current going through a wire, the force would total the result of the current and the magnetic field:

$$F = B \cdot I \cdot L \sin \theta$$

where
  L = length of the wire through the
      magnetic field (m)

24. Magnetic Field of a Moving Charge

The magnetic field near a long current-charge wire, in circular about the wire, is given by

$$B = \frac{\mu_0 I}{2\pi r}$$

where
  B = strength of the magnetic field (T)
  I = current through the wire (A)
  r = perpendicular distance from the center of the
      wire (m)

μ_0 = permeability of empty space

$\mu_0 = 4\pi \times 10^{-7}$ (H/m)

The henry (H) is the unit of inductance.

$$1\text{H} = 1\frac{\text{N} \cdot \text{s}^2 \cdot \text{m}}{\text{C}^2} = 1\frac{\text{Wb}}{\text{A}}$$

25. Magnetic Field of a Loop

For a long coil that is tightly turned, the magnetic field strength at its center is

$$B = \mu_0 I n$$

where

$n =$ number of turns per unit length of solenoid (turns/m)

$B =$ magnetic field in the region at the center of the solenoid (T)

$\mu_0 =$ permeability constant ($\mu_0 = 4\pi \times 10^{-7}$ H/m)

$I =$ current through the solenoid (A)

26. Faraday's Law

If the magnetic flux changes $d\Phi$ in a time dt, then the induced ε is given by

$$\varepsilon = -N\frac{d\Phi}{dt}$$

where

$\varepsilon =$ induced electromotive force (V)

$d\Phi =$ rate of change of the magnetic flux (Wb)

$dt =$ rate of change of the time (s)

$N =$ numbers of turns per loop.

(-) = the minus sign means that the magnetic field produced by the induced current opposes the external field produced by the magnet

27. Properties of Alternating Current

An alternating current (AC) is an electrical current in which the magnitude and direction of the current varies cyclically, as opposed to direct current, in which the direction of the current stays constant. The usual wave form of an AC power circuit is a sine wave, as this results in the most efficient transmission of energy.

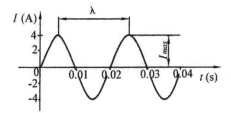

28. Period

The time required to complete one cycle of a waveform is called the period of the wave:

$$t = \frac{1}{f}$$

29. Frequency

The number of complete cycles of alternating current or voltage completed each second is referred to as the frequency:

$$f = \frac{1}{t}$$

30. Wavelength

The distance traveled by the sine wave during this period is referred to as the wavelength:

$$\lambda = \frac{c}{f}$$

where

c = speed of light $c = 3.00 \times 10^8$ (m/s)

31. Instantaneous Current and Voltage

Instantaneous current is the current at any instant of time. Instantaneous voltage is the voltage at any instant of time:

$$i = I_{max} \sin \theta, \quad e = E_{max} \sin \theta$$

where

i = instantaneous current (A)
I_{max} = maximum instantaneous current (A)
e = instantaneous voltage (V)
E_{max} = maximum instantaneous voltage (V)
θ = angle measured from beginning of cycle

32. Effective Current and Voltage

A direct measurement of AC is difficult because it is constantly changing. The most useful value of AC is based on its heating effect and is called its effective value. The effective value of an AC is the number of amperes that produce the same amount of heat in a resistance as an equal number of amperes of a steady direct current. The equations for effective current respectively voltage are

$$I_{eff} = 0.707 I_{max}$$
$$E_{eff} = 0707 E_{max}$$

where

I_{eff}, E_{eff} = effective value of current, and voltage

I_{max}, E_{max} = maximum or peak current, and voltage

33. Maximum Current and Voltage

When I_{eff} or E_{eff} is known, you can find I_{max} by using the formulas

$$I_{maf} = 1.41 I_{eff}$$
$$E_{max} = 1.41 E_{eff}$$

34. Ohm's Law of AC Current Containing Only Resistance

Many AC circuits contain resistance only. The rules for these circuits are the same rules that apply to DC circuits. The Ohm's Law formula for an AC circuit is

$$I = \frac{E}{R}$$

NOTE: Do not mix AC values. When you solve for effective values, all the values you use in the formula must be effective values.

35. AC Power

When AC circuits contain only resistance, power is found in the same way as in DC circuits

$$P = I^2 R = EI = \frac{E^2}{R}$$

36. Changing Voltage with Transformers

If we assume no power loss between primary and secondary coils, we have the following equation:

$$\frac{E_P}{E_S} = \frac{N_P}{N_S}$$

where

E_P = primary voltage (V)
E_S = secondary voltage (V)
N_P = number of turns in the primary coil
N_S = number of turns in the secondary coil

The relationship between primary and secondary current is

$$\frac{I_S}{I_P} = \frac{N_P}{N_S}$$

where

    I_S = current in secondary coil (A)
    I_P = current in primary coil (A)
    N_P = number of turns in primary
    N_S = number of turns in secondary

37. Inductive Reactance

The opposition to AC current flow in an inductor is called inductive reactance and is measured in ohms:

$$X_L = 2\pi fL$$

where

    X_L = inductive reactance (Ω)
    f = frequency of the AS voltage (Hz)
    L = inductance (H)

The current in a circuit that has only an AC voltage source and inductor is given by

$$I = \frac{E}{X_L}$$

where

    I = current (A)
    E = voltage (V)
    X_L = inductive reactance (H)

---

38. Inductance and Resistance in Series

The effect of both the resistance and the inductance on a circuit is called the impedance:

$$Z = \sqrt{R^2 + X_L^2} = \sqrt{R^2 + (2\pi fL)^2}$$

where

Z = impedance (Ω)
R = resistance (Ω)
XL = inductive reactance (Ω)
f = frequency of the AS voltage (Hz)
L = inductance (H)

a) Phase angle

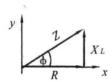

The phase angle is given by

$$\tan \phi = \frac{X_L}{R}$$

The resistance is always drawn as a vector pointing in the positive x-axis, and inductive reactance is drawn as a vector pointing into the positive y-axis.

b) Ohm's law

In general, Ohm's law cannot be applied to alternating-current circuits since it does not consider the reactance which is always present in such circuits:

$$I = \frac{E}{Z}$$

where

I = current (A)
Z = impedance (Ω)
E = voltage (V)

39. Capacitance

The effect of a capacitor on a circuit is inversely proportional to frequency and is measured as capacitive reactance, which is given by

$$X_C = \frac{1}{2\pi f C}$$

where

X_C = capacitive reactance (Ω)
f = frequency (Hz)
C = capacitance (F)

40. Capacitance and Resistance in a Series

The impedance of the circuit measures the combined effect of resistance and capacitance in a series

$$Z = \sqrt{R^2 + X_C^2} = \sqrt{R^2 + (2\pi fC)^2}$$

where

Z = impedance (Ω)

R = resistance (Ω)

X_C = inductive reactance (Ω)

f = frequency of the AC voltage (Hz)

C = capacitance (F)

a) Current

The formula for current is given by Ohm's law:

$$I = \frac{E}{Z}$$

where

I = current (A)

Z = impedance (Ω)

E = voltage (V)

b) Phase angle

The phase angle gives the amount by which the voltage lags behind the current:

$$\tan \phi = \frac{X_C}{R}$$

41. Capacitance, Inductance, and Resistance in Series

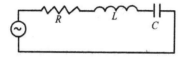

The impedance of a circuit containing resistance, capacity, and inductance in series can be calculated by the equation,

$$Z = \sqrt{R^2 + \left(X_L - X_C\right)^2}$$

where

Z = impedance (Ω)

R = resistance (Ω)

X_C = capacitive reactance (Ω)

X_L = inductive reactance (Ω)

a) Phase angle:
The phase angle is given by the following formula:

$$\tan \phi = \frac{X_L - X_C}{R}$$

b) Current

The current in this type of circuit is given by

$$I = \frac{E}{Z} = \frac{E}{\sqrt{R^2 + \left(2\pi fL - \dfrac{1}{2\pi fC}\right)^2}}$$

c) Frequency

The resonant frequency occurs when $X_L = X_C$. This frequency can be calculated by

$$f = \frac{1}{2\pi\sqrt{LC}}$$

42. Power in AC Circuits

When the current and voltage are in phase, then power can be stated as

$$P = EI$$

where

P = power (W)
E = voltage (V)
I = current (A)

a) Apparent Power

If current and voltage are not in phase, the resultant product of current and voltage is apparent power (S).

$$S = E \cdot I = \sqrt{P^2 + Q^2} = I^2 Z$$

b) Real power

Real power or actual power (P) is the product of apparent power (S) and the power factor:

$$P = E \cdot I \cdot p_f$$

c) Power factor:

$$p_f = \frac{P}{S}$$

where

p_f = power factor

P = real power (W)

S = apparent power (VA)

If φ is the phase angle between the current and voltage, then the power factor is equal to $\left|\cos\phi\right|$ and the real power is

$$P = S\cos\phi$$

d) Reactive Power

Reactive power (Q) is the power returned to the source by the reactive components of the circuit:

$$Q = I_L^2 X_l - I_C^2 X_C$$

where

$Q =$ reactive power (VAr)

$I_L =$ inductive current (A)

$I_C =$ capacitive current (A)

$X_L =$ inductive reactance (Ω)

$X_C =$ capacitive reactance (Ω)

43. Parallel Circuit

There is one major difference between a series circuit and a parallel circuit. The difference is that the current is the same in all parts of a series circuit, whereas voltage is the same across all branches of a parallel circuit. Because of this difference, the total impedance of a parallel circuit must be computed on the basis of the current in the circuit.

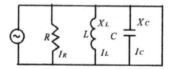

When working with a parallel circuit, one must use the following formulas:

a) Voltage

Voltage is the same across all branches of a parallel circuit. Thus,

Alternating Current

$$E = E_R = E_L = E_C$$

where

E = total voltage across circuit (V)

E_L = inductive voltage (V)

E_R = resistance voltage (V)

E_C = capacitive voltage (V)

b) Current:

$$I_Z = \sqrt{I_R^2 + I_X^2} = \sqrt{I_R^2 + \left(I_L - I_C\right)^2}$$

$$I_X = I_L - I_C.$$

where

I_Z = impedance current (A)

I_R = resistance current (A)

I_L = capacitive current (A)

c) Impedance

The impedance Z of a parallel circuit is found by the formula,

$$Z = \frac{E}{I_Z} = \frac{E}{\sqrt{I_R^2 + I_X^2}}$$

LIGHT

In a strict sense, light is the region of the electromagnetic spectrum that can be perceived by human vision, i.e., it is the visible spectrum, which includes wavelengths ranging approximately from 0.4 μm to 0.7 μm.

This section contains the most frequently used formulas, rules and definitions relating to the following:

1. General Terms
2. Photometry
3. Reflection, Refraction, Polarization
4. Geometric Optics

General Terms

---

1. Visible Light

Visible light is the portion of the electromagnetic spectrum between the frequencies of 3.8 10 ^{14} Hz and 7.5 10 ^{14} Hz. Hence,

$$3.8 \times 10^{14} \leq f \leq 7.5 \times 10^{14} \text{ (Hz)}$$

2. Speed of Light

The speed of light is a scalar quantity, having only magnitude but no direction. The following basic relationship exists for all electromagnetic waves, and relates the frequency, wavelength, and the speed of the waves. It is,

$$c = \lambda f$$

where

c = speed of light, 3.00×10^8 (m/s)
f = frequency (Hz)
λ = wavelength (m)

3. Light as a Particle

In quantum theory, particles of light are given the name "photons." A photon has energy defined by the equation,

$$E = hf = \frac{hc}{\lambda}$$

where

E = energy (J)

h = Planck's constant, $h = 6.62 \times 10^{-34}$ (J.s)

f = frequency (Hz)

λ = wavelength (m)

c = speed of light, 3.00×10^{8} (m/s)

4. Luminous Intensity

Luminous intensity refers to the amount of luminous flux emitted into a solid angle of space in a specified direction:

$$I_v = \frac{r^2 E_v}{\cos \theta}$$

where

I_v = luminous intensity (cd)

r = distance between the source and the surface (m)

E_v = illuminance (lux)

5. Luminous Flux

Luminous flux is a measure of the energy emitted by a light source in all directions:

$$\Phi_v = \Omega I_v$$

where

Φ_v = luminous flux (lm)

Photometry

Ω = solid angle (sr)

I_v = luminous intensity (cd)

6. Luminous Energy

Luminous energy is photometrically weighted radiant energy:

$$Q_v = \Phi_v t$$

where

Q_v = luminous energy (lms)

Φ_v = luminous flux (lm)

t = time (s)

7. Illuminance

Illuminance is the luminous flux collected by a unit of a surface:

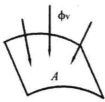

$$E_v = \frac{\Phi_v}{A} = \frac{\Omega I_v}{A}$$

where

E_v = illuminance (lx)

Φ_v = luminous flux (lm)

Ω = solid angle (sr)

I_v = luminous intensity (cd)

A = surface (m^2)

8. Luminance

Luminance is the luminous intensity emitted by the surface area of one square meter of the light source. The luminance value indicates glare and discomfort when we look at a lighting source. The following figure shows the concept:

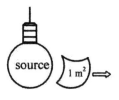

$$L_v = \frac{I_v}{S}$$

where

L_v = luminance $\left(cd/m^2\right)$

I_v = luminous intensity (cd)

S = area of the source surface perpendicular to the given direction $\left(m^2\right)$

9. Laws of Reflection

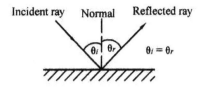

A ray of light is a line whose direction gives the direction of flow of radiant energy.

a) First law of reflection
The angle of incidence is equal to the angle of reflection. That is,

$$\theta_i = \theta_r$$

where

θ_i = angle of incidence

θr = angle of reflection

b) Second law of reflection
The incident ray, the reflected ray, and normal to the surface all lie in the same plane.

10. Refraction

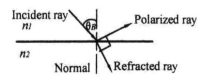

In an isotropic medium, rays are strength lines, along which energy travels at speed:

$$V = \frac{c}{n}$$

where

    n = refractive index of the medium
    c = speed of light in vacuum (m/s)

a) Law of refraction

When a ray of light passes at an angle from a medium of less optical density to a denser medium, the light ray is bent toward the normal.

When a ray of a light passes, at an angle, from a denser medium to one less dense, the light is bent away from the normal. Hence,

$$\frac{\sin \theta_i}{\sin \theta_2} = \frac{n_2}{n_1}$$

$$n_1 = \frac{c}{V_1}, \qquad n_2 = \frac{c}{V_2}, \qquad \frac{n_2}{n_1} = \frac{V_1}{V_2},$$

where

    V_1 = speed of light in a medium 1, (m/s)
    V_2 = speed of light in a medium 2, (m/s)
    n_1 = refractive index of the medium 1,
    n_2 = refractive index of the medium 2,
    c = speed of light in vacuum (m/s)

If $n_1 > n_2$ and θ_i exceeds the critical θ_c, where

$$\sin \theta_c = \frac{n_2}{n_1},$$

then there will be no refracted ray; this is a phenomenon called *total reflection*

11. Polarization

An electromagnetic or other transverse wave is polarized whenever the disturbance lacks cylindrical symmetry about the ray direction.

When the reflection is at 90^0 to the refraction, the transverse component of the electric field lies along the path of the reflection.

This would make the wave longitudinal, so clearly there is no transverse component in the reflection.

The incident angle at which this happens is called the polarizing angle or Brewster's angle:

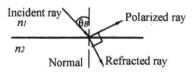

$$\tan \theta_B = \frac{n_2}{n_1}$$

where

θ_B = Brewster's angle $(^0)$

n_1 = refractive index of the incident medium

n_2 = refractive index of the reflecting medium

12. Plane Mirrors

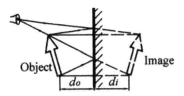

The image is at the same distance behind the mirror as the object is in front of it:

$$d_o = d_i$$

13. Concave Mirrors

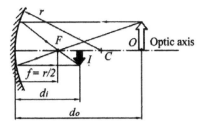

Depending upon the position of object, the image will be real or virtual.

14. Convex Mirrors
Convex mirrors produce only virtual and smaller images.

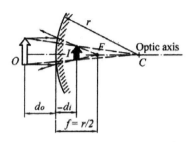

15. Mirror Formula

$$\frac{1}{f} = \frac{1}{d_o} + \frac{1}{d_i}; \qquad \frac{h_1}{h_o} = \frac{d_i}{d}$$

where

f = focal length of mirror
d_o = distance of object from mirror
d_i = distance of image from mirror
h_i = image height
h_o = object height

16. Lens Equation

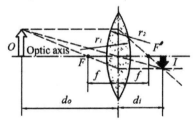

LIGHT
Geometrical Optics

$$\frac{1}{f} = \frac{1}{d_i} + \frac{1}{d_o} = (n-1)\left(\frac{1}{r_1} + \frac{1}{r_2}\right); \quad m = \frac{h_i}{h_o} = \frac{d_i}{d_o}$$

where

f = focal length

F, F' = focuses

r_1, r_2 = radii of curvatures

n = refractive index

h_i = image height

h_o = object height

m = magnification factor

d_o = object distance from lens center

d_i = image distance from lens center

WAVE MOTION AND SOUND

Wave motion is defined as the movement of a distortion of a material or medium, where the individual parts or elements of the material only move back and forth, up and down, or in a cyclical pattern.

This section contains the most frequently used formulas, rules, and definitions relating to the following:

1. Wave Terminology
2. Wave Phenomena
3. Electromagnetic Wave, Energy, and Spectrum
4. Sound Waves

---

1. Definition and Graph

A wave is a transfer of energy, in the form of a disturbance, through some medium, but without translocation of the medium.

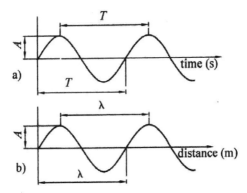

Waves may be graphed as a function of time (a) or function of distance (b) . A single frequency wave will appear as a sine wave in either case. From the distance graph, the wavelength may be determined. From the time graph, the period and frequency can be obtained. From both together, the wave speed can be determined.

2. Wavelength

Wavelength λ is defined as the distance from one crest (or maximum of the wave) to the next crest or maximum.

3. Amplitude

The amplitude A of a wave is the maximum displacement from the equilibrium or rest position.

4. Velocity

The velocity v of the wave is the measurement of how fast a crest is moving from a fixed point:

$$v = \frac{\lambda}{T} = \lambda f$$

where

v = velocity (m/s)
T = period (s)
f = frequency (1/s or Hz)
λ = wavelength (m)

5. Frequency

The frequency f of waves is the rate at which the crests or peaks pass a given point:

$$f = \frac{1}{T}$$

6. Period

The period T is the time required to complete one full cycle

$$T = \frac{1}{f}$$

---

7. Wave on a Stretched String

The speed of a wave traveling on a stretched uniform string is given by

$$V = \sqrt{\frac{F}{\rho}}$$

where

  F = tension in the string

  ρ = linear density of the string

8. The Sinusoidal Wave

The sinusoidal wave is a periodic wave described by a function of two variables of the form,

$$y(x,t) = A\cos\left[k(x - vt)\right]$$

where

  $y(x, t)$ = transverse displacement) at position x and time t

  A = amplitude

  k = angular wave number

  v = wave speed

a) Wave speed:

$$V = \frac{\varpi}{k}$$

b) Period

For a particular x, y is a periodic function of t with period:

$$T = \frac{2\pi}{\varpi}$$

c) Wavelength

For a particular t, function y is a periodic function of x, with the wavelength given by

$$\lambda = \frac{2\pi}{k}$$

d) Power

The average power transmitted by a sinusoidal wave can be calculated by the formula

$$P_{avg} = \frac{1}{2}\omega^2 A^2 \rho v$$

where

A = amplitude

ρ = density of a medium

v = wave speed.

ω = angular frequency

e) Energy

For a wave on string, the energy per unit length is given by

$$E_l = \frac{P_{avg}}{v}$$

where

P_{avg} = average power transmitted by the wave

v = wave speed

---

9. Electromagnetic Waves

These waves are made up of electric and magnetic fields whose strengths oscillate at the same frequency and phase. Unlike mechanical waves, which require a medium in order to transport their energy, electromagnetic waves are capable of traveling through a vacuum.

Although they seem different, radio waves, microwaves, x-rays, and even visible light are all waves of energy called electromagnetic waves.

Electromagnetic waves have amplitude, wavelength, velocity, and frequency. The creation and detection of the wave depend on the range of wavelengths.

a) Wave speed:

$$v = c = \lambda f = \frac{\lambda}{T}$$

where

c = speed of light (3.00×10^8 m/s)
f = frequency (1/s)
λ = wavelength (m)
T = period (s)

10. Electromagnetic Energy

Electromagnetic energy at a particular wavelength λ (in vacuum) has an associated frequency f and photon energy E:

$$E = h \cdot f$$

where

h = Planck's constant, $h = 6.62607 \times 10^{-34}$ (Js)
f = frequency (1/s)

11. The Electromagnetic Spectrum

The electromagnetic spectrum is a continuum of all electromagnetic waves arranged according to frequency and wavelength, as shown below

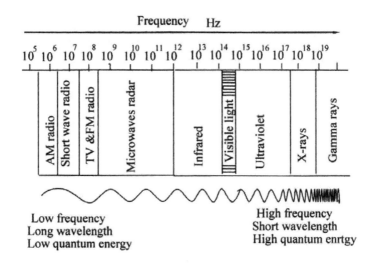

12. Sound Waves

Sound is a longitudinal wave in a medium created by the vibration of some object:

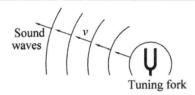

Tuning fork

13. Speed of Sound in Air

The speed in dry air at 1 atmosphere pressure and $0^0 C$ is 331.4 m/s. Changes in humidity and temperature cause a variation in the speed of sound. The speed of sound increases with temperature at the rate of 0.61 m/s $^0 C$. The speed of sound in dry air at 1 atmosphere pressure is then given by

$$v = 331.4 + (0.610) \cdot t_C$$

where

t_C = air temperature ($^0 C$)

14. Sound Speed in Gases

The speed of sound in an ideal gas is given by the formula

$$v = \sqrt{\frac{\gamma R T}{M}}$$

where

v = speed of sound (m/s)
R = universal gas constant = 8.314 J/mol K
T = absolute temperature (K)
M = molecular mass of gas (kg/mol)
γ = adiabatic constant

For air, the adiabatic constant $\gamma = 1.4$ and the average molecular mass (M) for dry air is 28.95 g/mol. Hence,

$$V = \sqrt{\frac{\gamma RT}{M}} = \sqrt{\frac{1.4(8.314)T}{0.02895}} = 20.05\sqrt{T} \text{ (m/s)}$$

15. The Doppler Effect

Suppose that a source emitting sound waves of frequency f_s and an observer move along the same straight line. Then the observer will hear sound of the frequency

$$f_0 = f_s \frac{v \pm v_0}{v \mp v_s}$$

where

> f_s = the source sound frequency
> f_0 = the observer sound frequency
> v_0 = the relative speed of the observer
> v_s = the relative speed of the source
> v = the sound speed in this medium

The choice of using a plus (+) or minus (-) sign is made according to the convention that if the source and observer are moving towards each other the observer frequency f_0 is higher than the actual frequency f_s. Likewise, if the source and observer are moving away from each other, the observer frequency f_0 is lower than the actual frequency f_s.

APPENDIX

Fundamental Physical Constants

Name	Symbol and Value
alpha particle mass	$m_\alpha = 6.6446565 \times 10^{-27}$ kg
atomic mass constant	$m_u = 1.66053886 \times 10^{-27}$ kg
atomic unit of energy	$E_h = 4.35974417 \times 10^{-18}$ J
atomic unit of force	$E_h/a_0 = 8.2387225 \times 10^{-8}$ N
atomic unit of length	$a_0 = 0.5291772108 \times 10^{-10}$ m
atomic unit of mass	$m_e = 9.1093826 \times 10^{-31}$ kg
Avogadro's constant	$N_A = 6.0221415 \times 10^{23}$ mol^{-1}
Bohr radius	$a_0 = 0.5291772108 \times 10^{-10}$ m
Boltzmann constant	$k_B = 1.3806505 \times 10^{-23}$ J K^{-1}
classical electron radius	$r_e = 2.817940325 \times 10^{-15}$ m
elementary charge	$e = 1.60217653 \times 10^{-19}$ C
electron charge to mass quotient	$\dfrac{-e}{m_e} = -1.758\,82012 \times 10^{11}$ C kg^{-1}

Continued

electron gyromagnetic ratio	$\gamma_e = 1.760\,85974 \times 10^{11}$ s^{-1} T^{-1}
electron magnetic moment	$\mu_e = -928.476412 \times 10^{-26}$ J T^{-1}
electron g factor	$g_e = -2.0023193043718$
Faraday's constant	$F = 96485.3383$ C mol^{-1}
fine-structure constant	$\alpha = 7.297352568 \times 10^{-3}$
molar mass constant	$M_u = 1 \times 10^{-3}$ kg mol^{-1}
molar volume of ideal gas	$V_m = 22.710981 \times 10^{-3}$ m^3 mol^{-1}
neutron g factor	$g_n = -3.82608546$
neutron gyromagnetic ratio	$\gamma_n = 1.83247183 \times 10^8$ s^{-1} T^{-1}
neutron mass	$m_n = 1.67492728 \times 10^{-27}$ kg
Newtonian constant of gravitation	$G = 6.6742 \times 10^{-11}$ m^3 kg^{-1} s^{-2}
nuclear magneton	$\mu_N = 5.05078343 \times 10^{-27}$ J T^{-1}
Planck's constant	$h = 6.6260693 \times 10^{-34}$ J s
Planck mass	$m_P = 2.17645 \times 10^{-8}$ kg

Continued

proton charge to mass quotient	$\dfrac{e}{m_p} = 9.57883376 \times 10^7$ C kg^{-1}
proton g factor	$g_p = 5.585\ 694701$
proton gyromagnetic ratio	$\gamma_p = 2.67522205 \times 10^8$ s^{-1} T^{-1}
proton mass	$m_p = 1.67262171 \times 10^{-27}$ kg
proton-electron mass ratio	$\dfrac{m_p}{m_e} = 1836.15267261$
speed of light in vacuum	$c = 299792458$ m s^{-1}
standard acceleration of gravity	$g = 9.80665$ m s^{-2}
standard atmosphere	$p = 101325$ Pa
Stefan-Boltzmann constant	$\sigma = 5.670400 \times 10^{-8}$ W m^{-2} K^{-4}

absolute value, 32
AC power, 316
acceleration, 227, 242, 244
acute angl, 75
adding and subtracting solynomials, 39
addition propertie, 28
alternate exterior angles, 78
alternate interior angles, 77
altitude, 88, 93,
amortization of loans, 174
ampere, 3
angle, 75
angle between two lines, 151
angle bisector, 78
angled projection, 240
angular acceleration, 229
angular displacemen, 228
angular measure, 117

angular velocity, 228
annuities, 171
annulus, 98
apparent power, 322
arc length, 200
Archimedes' Principle, 272
area and volume expansion:, 284
arithmetic mean, 57
arithmetic sequenc, 56
arithmetic series, 57
associative, 28
Avogadro's number, 281
barrel, 113
base units, 3
Bernoulli's equation, 275
binomial coefficient., 61
binomial expansion, 67
binomial theorem, 61
Boltzmann's constant, 286
brackets, 38

capacitance, 319
capacitances in series, 300
Capacitor, 300
Cartesian coordinate system, 63
Celsius, 279
center of gravit, 137
central angles, 95
centrifugal force, 250
centripetal acceleration, 232
centripetal force, 251
chain rule, 182
Charles Law, 287
circle, 95, 141
circular and angular measure, 117
circular ring, 98
circumscribed circl, 80
cirresponding angles, 77
closed interva, 31
Combinations, 212
combined gas law, 289
common logarithm, 71
commutative, 28
complementary angles, 76

complex fraction, 43
complex number, 44
compound interes, 168
cone, 103
conjugate of a complex number, 45
conservation of energy, 255
continuous compound interest, 169
conversion period, 168
corresponding angles, 78
Coulomb's Law, 295
cube, 100
cuboid, 101
current density, 302
cylinder, 106
cylindrical surfac, 163
De Morgan's Laws, 210
definition of an angle, 75
denominator., 51
density, 267
derivative, 181
determinants, 49
diagonal of, 100
diagonals, 90, 92,

direction of a line, 151
discriminant, 54, 66
displacement, 227
distance, 227
distance between two
    points, 133
distance from a line,
    140
distributive, 28
dividing polynomials,
    39
division, 29
Doppler Effect, 348
effective current, 314
effective rate, 169
electric fux, 296
electric potential, 298
electromagnetic energy,
    345
electromagnetic
    spectrum, 346
electromagnetic waves,
    345
ellipse, 98
ellipsoid, 112, 159
elliptic paraboloid, 160
empty set, 207
energy, 254, 301

equal length, 87
equality, 28
equation of a circle, 142
equilateral triangle., 86
equivalence, 29
expansion of water, 285
exponent, 33
exponential distribution,
    219
exponential function.,
    67
external angles, 95
factoring, 52
factoring a polynomial,
    36
factorization:, 124
Fahrenheit, 279
failure distribution
    function, 220
failure rate, 220
Faraday's Law, 312
First Law of
    Thermodynamics,
    289
flow rate, 274
foci, 98
force on a horizontal
    surface, 270

force on a vertical surface, 270
forms of linear equations, 64
fractions of, 29
frequency, 228
friction, 243, 245
frustum of a con, 105
frustum of a pyramid, 103
future value, 167
future value of an annuity, 173
Gauss' Law, 289
Gay-Lussac's Law, 288
geocenter, 81
geometric mean, 59
geometric sequence, 58
gravitational force, 250
half-closed interval, 32
half-open interval, 32
hertz, 4
hollow cylinder, 107
horizontal projection, 241
hyperbola, 146
hyperbolic paraboloid, 161

hyperboloid, 160
hyperboloid of two sheets:, 161
identity, 28
illuminance, 331
imaginary part, 44
impedance, 325
impulse of force, 247
inclination, 135
inclination and slope of line, 134
inclinedplane, 261
increasing and decreasing function, 186
inductive reactance, 317
inequalities, 33, 56
Initial speed, 239
Inscribed circle, 80, 87
instantaneous, 314
integration by substitution, 189
integration rules, 188
intercept form, 65
interest rate, 167
internal angles, 95
internal resistance, 309
International System, 3

interval, 31
inverse, 28
irrational numbers, 27
isosceles triangle, 87
Joule's Law, 306
kelvin, 3,
Kelvin, 279
kilogram, 3
kinetic energy, 254
Kirchhoff's Junction
    Law, 307
kite, 93
lateral area, 104, 106
law of cosines, 82
law of cosines, 83
law of gravity, 249
laws of logarithms, 70
laws of reflection, 333
LCD, 42
least common
    denominator, 42
length, 3
length of arc, 97
lens equation, 337
lever, 259
light as a particle, 329
limits, 179
linear equation, 47,69

linear expansion, 281
logarithmic function.,
    69
lorentz factor, 256
luminous energy, 331
luminous intensity, 330
magnetic field of, 311
magnetic forces on, 310
mass, 3
maximum curren, 315
mean time to failure,
    221
meter, 3
method of elimination,
    51
method of substitution:,
    48
mirrors, 336
molar mass, 281
molar volume, 282
mole, 3
moment of a force, 258
multiplication
    properties, 28
multiplying
    polynomials, 39
natural exponential
    function, 68

natural exponential function., 68
natural logarithm, 71
newton, 5
Newton's First Law, 246
Newton's Second Law, 246
Newton's Third Law, 246
nominal interest rate, 170
normal distribution, 218
normal form of equation, 139
number of diagonals, 95
numerator, 46
oblique triangle, 79
obtuse angle, 76
Ohm's Law, 303
open interva, 31
operation with rational expressions, 41
operations with complex numbers, 44
origin, 63
orthocenter, 81
parabola, , 145

parallel circuits, 304
parallelogram, 90
parametric form of, 182
parenthesis, 38
pendulum, 237
perimeter, 99
period, 228, 238, 313
permutation, 211
perpendicular angle, 79
phase angle, 318
point of division, 133, 150
Poisson Distribution, 218
polar coordinates, 148
polarization, 335
polynomial of, 39
positive integer, 33
potential energy, 255
power, 35
power factor, 323
powers of i, 45
present value, 170
present value of annuit, 172
pressure, 279
pressure in liquid, 269
principa, 167

probability, 213
properties of absolute
  value, 32
properties of equality,
  28
properties of rational
  expressions, 41
properties of the
  exponential function,
  68
properties of the
  logarithmic
  functions, 70
pulley, 260
pyramid, 102
Pythagoream theorem,
  85
quadratic equations, 51
quadratic function, 65
radian, 4, 75
radical, 37
radicals, 35
random variable, 214
Rankine, 279
rational expression, 40
rational number, 27
rays, 75
reactive power, 323

real number line, 30
real numbers, 27
real power, 323
rectangle, 89
reduce to base
  equation:, 124
regular polygon, 94
relativistic rnergy, 256
reliability function, 220
reliability of the system,
  221
remainder, 45
resistance, 302
resistors, 319
rhombus, 91
right angl, 76
right prism, 101
right triangle, 85
rolling motion, 243
rotational motion, 232
rotational speed, 229
scalars, 227
screw, 263
secant, 97
Second Law of
  Thermodynamics,
  290
sector of a circle, 96

sector of a sphere, 110
segment of, 97
segment of a circle, 97
series crcuits, 304
series-parallel circuit, 305
set and notation, 207
set equality, 207
set of, 27
set union, 208
sign, 29
similarity of triangle, 81
simple harmonic motion, 235
simpleinterest, 167
sinking fund payment, 174
slant height, 104,106
sliced cylinder, 108
sliding motion, 242
slop, 135
slope intercept form, 64
sope of tangent line, 180
solid angle, 332
solving quadratic equations by factoring, 52

specific heat, 283
specific volume, 280
speed, 228
speed of ligh, 329
speed of Sound in Air, 347
sphere, 109
spherical cap, 109
square, 89
square of a binomial, 37
standard value, 217
steradian, 4
subset, 207
substitution, 29
sum of the angles in a triangle, 81
supplementary angle, 76
systems of linear equations, 47
tangential acceleration:, 232
temperatur, 279
Thales' theorem, 79
Third Law of Thermodynamics, 292
tilting angl, 244
time, 3

torque, 252
torus, 111
trajectory velocit, 241
transformers, 316
transitive, 28
trapezoid, 92
trigonometric circle, 118
trigonometric equations, 124
trigonometric identities, 126
two-point form, 65
U.S. units, 12
uniform accelerated, 230
uniformlinear motion, 229
vaporization, 286
variance, 215
vectors, 227
velocity, 228
Venn diagrams, 208
vertex, 75
vertical angle, 77
Vieta's rule, 52
viscosity, 267
wavelength, 341
wedge, 262
work, 252
zone of a sphere, 110